Solar Technology and Global Environmental Justice

Building on insights from ecological economics and philosophy of technology, this book offers a novel, interdisciplinary approach to understand the contradictory nature of solar photovoltaic (PV) technology.

PV technology is rapidly emerging as a cost-effective option in the world economy. However, reports about miserable working conditions, environmentally deleterious mineral extraction and toxic waste dumps corrode the image of a problem-free future based on solar power. Against this backdrop, Andreas Roos explores whether 'ecologically unequal exchange' – an asymmetric transfer of labor time and natural resources – is a necessary condition for solar PV development. He demonstrates how the massive increase in solar PV installation over recent years would not have been possible without significant wage/price differences in the world economy – notably between Europe/North America and Asia – and concludes that solar PV development is currently contingent on environmental injustices in the world economy. As a solution, Roos argues that solar technology is best coupled with strategies for degrowth and counter-metabolic alliances, which allow for a transition away from fossil fuels and toward a socially just and ecologically sustainable future.

This book will be of great interest to students and scholars of solar power, philosophy of technology, and environmental justice.

Andreas Roos is an interdisciplinary scholar with a doctoral degree in the field of human ecology. His work draws from ecological economics, environmental history, and philosophy of technology to understand the contentious relation between technology and ecology. Roos's most recent work focuses on assessing the potential of renewable energy technologies to transform modern human-environmental relations. Publishing in top-ranking journals, Roos's other contributions include ecological perspectives on the digital economy and the possibilities for commons-based energy technology.

Routledge Studies in Environmental Justice

This series is theoretically and geographically broad in scope, seeking to explore the emerging debates, controversies, and practical solutions within Environmental Justice, from around the globe. It offers cutting-edge perspectives at both a local and global scale, engaging with topics such as climate justice, water governance, air pollution, waste management, environmental crime, and the various intersections of the field with related disciplines.

The *Routledge Studies in Environmental Justice* series welcomes submissions that combine strong academic theory with practical applications, and as such is relevant to a global readership of students, researchers, policy-makers, practitioners, and activists.

Diversity and Inclusion in Environmentalism
Edited by Karen Bell

Environmental Justice in the Anthropocene
From (Un)Just Presents to Just Futures
Edited by Stacia Ryder, Kathryn Powlen, Melinda Laituri, Stephanie A. Malin, Joshua Sbicca and Dimitris Stevis

John Rawls and Environmental Justice
Implementing a Sustainable and Socially Just Future
John Töns

Intergenerational Challenges and Climate Justice
Setting the Scope of Our Obligations
Livia Ester Luzzatto

Solar Technology and Global Environmental Justice
The Vision and the Reality
Andreas Roos

For more information about this series, please visit: www.routledge.com/Routledge-Studies-in-Environmental-Justice/book-series/EJS

Solar Technology and Global Environmental Justice

The Vision and the Reality

Andreas Roos

Routledge
Taylor & Francis Group

LONDON AND NEW YORK

earthscan
from Routledge

First published 2023
by Routledge
4 Park Square, Milton Park, Abingdon, Oxon OX14 4RN

and by Routledge
605 Third Avenue, New York, NY 10158

Routledge is an imprint of the Taylor & Francis Group, an informa business

British Library Cataloguing-in-Publication Data
A catalogue record for this book is available from the British Library

ISBN: 978-1-032-27338-9 (hbk)
ISBN: 978-1-032-27339-6 (pbk)
ISBN: 978-1-003-29231-9 (ebk)

DOI: 10.4324/9781003292319

Typeset in Goudy
by codeMantra

Contents

Acknowledgments

I wrote this book with the conviction that human societies cannot survive without healthy relations to their environments. By the same line of reasoning, there can be no inside without an outside; no us without a them; and no me without a you. I believe that our thoughts are rarely our own, or, to be more humble, that my thoughts are rarely my own. It is abundantly clear to me that I could not have written this book without the support of family, friends, and colleagues who shaped my thinking and supported me in many different ways.

I want to thank my family for always supporting me. Your love and your tireless encouragement has made me who I am today. It seems that you have always trusted me to find my own way and thereby taught me the subtle art of following the heart. Your support throughout the journey of writing this book has been nothing short of fantastic.

This book would not have been possible without the excellent support of my two PhD supervisors Alf Hornborg and Vasna Ramasar at the Human Ecology Division, Lund University. Your support sometimes extended well beyond your obligations, including encouraging me to rewrite my dissertation as a book, the one you are now reading. I am very grateful to you and to everyone at the Human Ecology Division.

While this book marks an end to my PhD studies, it also marks a new beginning. I am very grateful for the collaborations and friendships that are now forming with colleagues on the HARNESS project and at the Division of History of Science, Technology and Environment at KTH. Thank you for your kind support and encouragement.

My gratitude also goes to Marlis Wullenkord for your loving support and inspiration. It is a rare thing to have found someone who so passionately share my desire for a socially just and ecologically sustainable future and take active steps toward realizing it.

Lastly, I want to thank Robbie Smith, Emma Brolén, and her mother Helena for hosting me in the pastoral idyll of Sörmland while finishing the book. The last chapter – so different from the rest of the book – is laden with the prospects of active hope that only the direct experience of being with friends and nature, foraging, swimming, and creating music can inspire.

1 Solar technology at the brink

Solar photovoltaic (PV) technology is today one of the most favored responses to the ecological crisis. Simultaneously, researchers and environmentalists are concerned that the social and ecological consequences of solar power may be both harmful and unequally distributed (e.g., Zehner, 2012; Trainer, 2014; Yenneti et al., 2016). This concern has been met by polarized responses ranging from a rejection of solar PV technology (e.g., Zehner, 2012) to prompt solutions (e.g., Mulvaney, 2019) to outright denial (e.g., Phillips, 2020). The question whether the harmful and unequally distributed consequences are inherent or transitory to solar PV technology lies at the very heart of the polarization. Meanwhile, few are asking the most fundamental question of what solar PV technology *is*. In this book, I combine perspectives from philosophy of technology and ecological economics in the hope to provide an answer to this question.

The aim is (a) to understand whether an uneven distribution of resources in the world economy is inherent to large-scale solar PV development, (b) what this condition means for the definition of solar PV technology, and (c) what this in turn implies for the premise that solar PV technology is a feasible option for a socially just and ecologically sustainable world.[1] The purpose is to present propositions about ecologically unequal exchange and the ontology of technology that will clarify and deepen current discussions on solar power in a way that will be useful for the manifold efforts to create a socially just and ecologically sustainable world.

The book builds on the conviction that the conventional understanding of technology as "applied science" is fragmentary and misleading (e.g., Ellul, 1964; Heidegger, 1977[1954]; Winner, 1978, 2020[1986]; Feenberg, 1991; Marx, 2010). The typical problem with technological solutions is that the full social, political, and ecological conditions of their deployment become apparent only after significant changes have already occurred (Winner, 2020[1986]). "[A]s technologies are built and put into use," Winner suggests, "significant alterations in patterns of human activity and human institutions are already taking place. New worlds are being made" (Winner, 2020[1986]: 11). However, we seldom stop to consider what kind of world is already being made. This applies to solar PV technology as a contestant to fossil fuels. This time, however, the stakes are so consequential to the fate of the biosphere that we cannot afford to ignore what world they encourage.

DOI: 10.4324/9781003292319-1

Considering that solar power is no longer alternative technology, it is both possible and incumbent upon us to consider how visions of a solar-powered future compare to the social-ecological reality of its existence. There is a substantial amount of literature showing that the technologies that are applied to solve the ecological crisis seldom live up to their promises (e.g., Huesemann and Huesemann, 2011; Bonds and Downey, 2012; Giampietro and Mayumi, 2015). In many of these examples, including biofuels, electric cars, nuclear power, and carbon capture storage, governments, corporations, and developers pursue "green" technologies because they have been cloaked in a symbolism overshadowing the reality of their application.[2] Following these studies, the hypothesis of this book is that the conception of solar PV technology contradicts the actual conditions of its existence.

Human ecology, grounded in both the social and natural sciences, is uniquely equipped to understand how the symbolic meaning of solar PV technology relates to the physical conditions of its existence. As Paul Shepard (1967) contended long ago, "the central problem of human ecology may be characterized as the relationship of the mind to nature." I draw upon the human-ecological theory of "ecologically unequal exchange," which explains the mechanism by which industrial societies enjoy high levels of material wealth and technological development, while exporting the environmental loads of their lifestyles to the world's impoverished (see Chapter 2). Importantly, it has been argued that the rise of fossil-powered machinery and industrial manufacturing was contingent upon an ecologically unequal exchange whereby the British Empire appropriated large amounts of resources from North America through international trade (Hornborg, 2006).[3] So far, however, only two studies have empirically examined the relation between "green" technology and ecologically unequal exchange (Bonds and Downey, 2012; Hornborg et al., 2019).

The possibility that solar PV technology *requires* ecologically unequal exchange means that an uneven distribution of resources and environmental impacts in the world economy may be an overshadowed condition of solar PV technology. If this is so, solar PV technology is more than an "applied science" to harness the forces of nature, since it then involves globally uneven relations without which it would not exist. This raises important questions concerning the premise that application of solar PV technology inevitably leads to a more democratic, just, and environmentally sustainable world (a popular argument that I will soon examine). The purpose of this inquiry is not to portray solar PV technology as "bad," but to understand what type of relations solar proponents are *already* creating under the labels of "green" technology and "renewable" energy. The purpose is also to offer an alternative vision with practical solutions that go beyond these labels and the conventional narrative of emancipatory technology.

In drawing on insights and discussions from philosophy of technology, I situate the core questions of this book as part of a long tradition attempting to understand the inconsistencies and contradictions of technological solutions in industrial societies. Among philosophers such as Karl Marx, Lewis Mumford, Martin Heidegger, Jacques Ellul, Herbert Marcuse, Ivan Illich, Langdon Winner, and

Ernst Schumacher, there is a common ambition to examine the wider historical implication of advanced technology in industrial societies and its emancipatory promises. Importantly, these examinations were often raised from observations on how new machinery and new technologies were repeatedly inconsistent with their liberating and cornucopian promises. I draw upon these thinkers today, because it is now possible to identify contradictions in the visions of a future society based on solar PV technology.

The book focuses on the question of whether solar PV technology requires ecologically unequal exchange in the world economy. This means that the book does not provide a systematic overview of all concerns relevant for understanding solar technology and global environmental justice. As such, I have chosen to approach the question of solar power and ecologically unequal exchange as central for understanding the deeper structures underlying recent concerns over global environmental injustices. Rather than detailing the injustices in the global solar commodity chain, this book provides the reader with a comprehensive philosophical–historical framework and an empirically grounded analysis for understanding the present and future roles of solar PV technology. The empirical cases, focusing on Germany, China, the US, India, and Italy, reflect the ambition to understand the presence of ecologically unequal exchange among the most powerful pro-solar nations today. Considering the global character of the PV commodity chain, this focus will include considerations of key regions for the global solar industry (e.g., Southeast Asia and northern Europe). The choice of methodology – including philosophical inquiry, historical analysis, life cycle analysis-based accounting of ecologically unequal exchange, and calculations of power density and energy return on energy investment – reflects the ambition to generate a genuine interdisciplinary understanding of solar PV technology rooted in philosophical materialism.

In this chapter, I first examine the social-ecological arguments for solar power and how the visit to a solar park in Sweden compelled me to ask questions about these arguments. I then show how solar power is a celebrated technological solution across the political spectrum, all while there is a growing group of contenders pointing out its drawbacks and contradictions. This is followed by a note on the book's interdisciplinary approach. Finally, I provide a brief overview of the contents and structure of the book.

Relating solar visions to reality

Around 10–20 years ago, every other book on the state of the planet seemed to end with a hopeful note on some of the most promising technological solutions. The "good news," one well-known student textbook claimed, "is that we know a number of ways to slow the rate and degree of global warming and the resulting climate change caused by our activities" (Miller and Spoolman, 2009: 514). The solutions came down to three strategies: "(1) improve energy efficiency to reduce fossil fuel use; (2) shift from nonrenewable carbon-based fossil fuels to a mix of carbon-free renewable energy resources; and (3) stop cutting down tropical forests" (ibid.). The chapter proceeded to discuss 17 different solutions, out of which over a half implied either a novel application or significant development of an advanced technology. Among these solutions was the suggestion to "[i]ncrease solar power 700-fold to displace coal-fired plants," a figure that was much higher than the suggested installation of other renewable energy technologies (ibid.). At the time, countless books ended with similar visions proclaiming devotion to technological solutions and in particular solar power. Since then, much has changed, and it is now possible to assess the practical implications of these visions.

The predisposition toward technological solutions such as solar power is based on important *ecological* reasons. There is no doubt that fossil-based energy systems need to be replaced as swiftly as possible for world society to have any chance of mitigating anthropogenic emissions of greenhouse gases and stay below 2°C warming compared to pre-industrial levels (IPCC, 2014, 2018). Many are convinced that solar power will largely replace fossil energy. Unlike fossil energy, solar power does not directly emit greenhouse gases in the process of generating electricity, which means that it does not directly cause air pollution that adversely affects plants, humans, and non-human animals in local ecosystems. By all practical accounts, direct sunshine is also a near-infinite energy resource that does not disappear from the world once it is "used up," as is the case with the combustion of coal or oil.

There are also important *social* reasons for turning to solar power. Solar advocates have long pointed out how generating electricity from locally sourced direct sunshine is much more aligned with democratic values and just social relations (Mander, 1991; Scheer, 2007; Klein, 2014). Even technology critic Jerry Mander (1992), who accused the environmental movement in the 1990s for "failing to effectively criticize technical evolution despite its obvious, growing, and inherent bias against nature," saw solar power as "intrinsically biased toward democratic use" (Mander, 1992: 3, 36). This, he argued, was because solar technology "is buildable and operable by small groups, even families" and "does not require centralized control … a reason why big power companies oppose it" (ibid. 36).

Together, the combined social and ecological reasons for advocating solar power form a powerful foundation for contemporary visions of a more democratic and ecologically sustainable future. One influential vision can be found in Naomi Klein's (2014) acclaimed book *This Changes Everything*, in which she tells the story of Henry Red Cloud, a Lakota entrepreneur who is working with the local community in the Pine Ridge Reservation in South Dakota to install solar power. For Red Cloud, "'solar power was always part of Native's way of life'" because it embodies a worldview in which people are adapting to the rhythms of natural ecosystems, as opposed to dominating and controlling them (ibid. 393). The relinquishing of control, Klein continues, is inherent to solar power because in contrast to the power of coal, the sun does not shine all the time and cannot therefore generate energy on demand. This means that solar power also embodies fundamentally different "power relations between humanity and the natural world" and that humans have to work "with the earth" instead of simply consuming it, as if it was a resource (ibid. 239).

Klein argues that solar power should be understood as a tangible alternative for local communities to convert fossil infrastructure in their own backyard into locally owned and democratically operated solar parks. With examples from Navajo- and Hopi-led initiatives to convert a local coal mine into a solar park, she argues that these parks could provide training skills, jobs, and steady revenues for members in the community, all while generating "power not just for their reservation but also large urban centers" (ibid. 296). They would make a "fundamentally new [social-ecological] relationship" possible, in which communities themselves can set the framework and agenda for their own development while rejecting the extractive logic of fossil technology that destroys the land and the Earth (ibid. 297). In the case of the Navajo and Hopi, however, this vision could never be fully realized because of one major barrier: lack of capital.

Despite the absence of practical examples, the core thrust of Klein's vision remains strong: solar technology can inaugurate new social-ecological relations naturally aligned with ecological sustainability and democracy. For several reasons, rivaling energy options such as nuclear power cannot serve the same democratic, nature-bound basis for future societies. Nuclear power plants require expert knowledge on how to stabilize the fission process in the reactor core as well as expertize on how to build and maintain the necessary infrastructure and how to (not) take care of the radioactive waste materials. This immediately places a disproportionate amount of power – political and physical – in the hands of a few select experts. Contrary to capturing sunshine, the process of splitting atoms (fission) also implies that practitioners intentionally encourage and exploit a natural weakness (molecular instability) to destroy the very fabric of life. This is often done under the misleading narrative that it will be possible to contain the fission process within the reactor cores and avoid releasing it upon the world through accidental meltdowns or purposeful military operations (Sovacool, 2011; Wealer et al., 2019). Despite the ecomodernist conviction that "nuclear fission, and nuclear fusion represent the most plausible pathways towards the joint goals of climate stabilization and radical decoupling human from nature" (Asafu-Adjaye et al., 2015: 23–24), nuclear power is associated with higher demand on the natural world than fossil fuels (Diaz-Maurin and Giampietro, 2013). As such, solar power will likely remain a far more democratic and environmentally favorable option than nuclear power.

Armed with these arguments for solar power, I was curious to visit a local solar park that was founded upon values of ecological sustainability and grassroots democracy. In Sweden, the most famous solar park resembling what Klein described is developed and operated by a company called ETC just outside the town of Katrineholm. ETC is primarily a media company running an online media platform and publishing a monthly magazine consisting of left-leaning news that also caters to environmental concerns. Over the years, the company has gained some attention and is now making a name for itself among progressives and environmentalists. In contrast to the Navajo- and Hopi-led organization for solar power, ETC has been able to invest capital in a self-operated solar park funded by its readers, who either contribute financially through the subscription fee or invest directly in the solar park with an interest rate of 2%. ETC's solar park displays both similarities with and differences from the proposed solar park of the Navajo and Hopi. Most importantly, ETC seems to have solved the problem of capital.

I decided to contact the head of the park and was happy to receive the reply that I was more than welcome for a guided tour and a chat. A few months later, I was on my way to the park located roughly 2 km north of Katrineholm. I arrived by train and decided to walk to the solar park. To my surprise, arriving at the park by foot was rather difficult. My immediate thought was that the park must have been newly built and that they simply had not had time to construct a proper entrance. It was only later that I realized that the park entrance was located on the other side of the compound and that visitors were expected to access the solar park by car. After orienting myself from the backdoor, I could see that the park was located just next to a large road and that some of the constructions in the park (e.g., a large solar tower with the ETC logo) were meant as advertisements for cars passing by on the highway. The idea, I figured, was that the visitors coming by car were exposed to an association between solar power and electric vehicles as a viable transportation alternative for the future. I later learned that the park offers free electric charging for those who pass by with electric vehicles. In this sense, the solar park was not a solar park at all, but a gas station or, rather, a solar station. This came as a surprise to me, who had taken to heart Ivan Illich's conviction that "[p]articipatory democracy demands low energy technology, and free people must travel the road to productive social relations at the speed of a bicycle" (1973: 12).

At the park, the head of operations, David Eskilsson,[4] greeted me. He was just finishing an education session with a group of schoolchildren. David had been a project manager at the ETC solar park for over seven years and now thought of himself primarily as an inspirer for solar cells and climate transformation. It became apparent through our talk that the solar park was founded on the idea of demonstrating that solar PV cells and other solar power solutions are effective to combat climate change, regardless of what critics say. But technological solutions, David told me, were by no means the most important aspect of what they were doing. Instead, their primary objective was to demonstrate that alternative ways of harnessing energy are a feasible option for the future. The purpose was to make people feel empowered to install solar cells or solar modules of their own after a visit to the park.

To say that solar power is easy, David explained, is not to say that the development of solar projects is without setbacks. Technically, solar PV modules are easy to install, but the social-political context is not exactly generous to organizations (such as ETC) who are seeking to transform the nation's energy regime. He proceeded to give a brief account of the power dynamics between the electric grid operators, the role of the state, and how the park could be part of a grassroots movement that could influence political decisions, if it grew big enough. He pointed out that this had in fact already occurred in Germany, where small-scale energy producers together formed a political force of consequence. In 2010, I later learned, up to 40% of Germany's installed renewable energy capacity was owned by citizen energy cooperatives, so-called *Energiegenossenschaften* (Buchan, 2012: 9–11). Recently, David continued, this had become a tangible option for people in Sweden, because the price of solar energy had fallen considerably over the last years. This meant that energy cooperatives could soon wield considerable political power and influence many important decisions in Sweden, but only if people felt encouraged to install solar modules on their rooftops and organize into cooperatives.

When I asked why the prices had fallen, David told me that it was because of lower production costs in solar PV cell manufacturing. In turn, I asked why he thought that production costs of solar cell had dropped in prices over the recent years. "It is because of the Chinese," he replied without hesitation, "and that there is now a mass production [of solar PV modules]." We briefly discussed the issue of being dependent on the international market for solar PV modules. David thought that it would be difficult if Sweden sought to become completely self-sufficient in solar PV module production. A certain level of international trade would remain necessary, he asserted. With the analogy of textile production in India and Pakistan, he explained how clothes would be far too expensive for Swedish consumers if they were all produced in Sweden where labor is more expensive and where there are strong environmental regulations in place. "We need to find a good form for it [production of solar PV modules]," he argued, because we cannot completely get rid of this situation (ibid.). Both David and I saw the problem.

The problem was that the vision for an environmentally sustainable and democratically aligned solar-powered energy regime orchestrated by grassroots actors was founded upon what seemed to be a socially and environmentally dubious division of labor in the world economy. To keep prices on solar panels low enough to be politically subversive appeared to *require* that they be produced on the other side of the world at low wages, with low environmental regulations, and often in fossil-powered industries. This confirmed my suspicion that solar power, however decentralized it appears when installed on rooftops and balconies, is not a local affair at all. To the extent that solar PV modules are made within an economic system largely organized in terms of price differences, even the smallest local initiatives to put solar modules on rooftops is contingent on a wider global economy. While the powerful solar visions of Naomi Klein and ETC are ecologically and democratically aligned, neither of them fully account for the processes occurring outside the boundaries of the solar parks, where these price differences are exploited (even if David was certainly aware of the issue).

This means that the political visions for a solar-powered future may not be consistent with the reality of their practical implementation. To democratize the global commodity chains by increasing salaries and enforcing environmental regulations in China would not necessarily dissolve the inconsistency, because the solar parks and democratic solar cooperatives would then likely become unaffordable to Swedish citizens. The democratic possibility and sustainability potential of solar power available to the globally affluent may therefore be inextricably bound to the uneven and environmentally degrading conditions among the globally impoverished. This would mean that the massive push for solar PV technology is not only generating solar power in the sense of renewable energy, but also renewing the global social relations of power that has characterized the world economy at least since the 16th century. By extension, if solar power is being posited as one of the most promising technological solutions to environmental problems while simultaneously embodying notable environmental disadvantages and global asymmetries, have we really understood the nature of technological solutions to environmental problems, and, by extension, technology itself?

Ecomodernism vs. ecorealism

Despite disagreement among researchers and environmentalists, solar power is widely celebrated across the political spectrum. Over the last two decades, corporations from a diverse range of industries (oil to cosmetics) have transformed the production of solar technology into a profitable commercial industry spurred by capital accumulation (Nemet, 2019). This is a still ongoing process occurring with considerable support from socialist and neoliberal governments alike. The idea of a global "solar communism," according to which solar technology can be installed to overthrow capitalist relations of production is promoted by some eco-socialists (Schwartzman, 1996; Klein, 2014). In this camp, solar technology is understood as antithetical to corporate energy generation and the fossil-based transportation system that is at the heart of global capitalism (Malm, 2016). Some advocates of degrowth – a controlled downscaling of modern industrial societies to ecologically sustainable levels of matter-energy throughput under socially favorable conditions – also consider solar power as a promising solution, given the right circumstances (for an overview, see Dale, 2019). On the opposite side of the political spectrum, parties on the far right, particularly in some European countries such as France, have articulated ambitions to develop their countries' renewable energy capacity (Jeffries, 2017). This demonstrates how the aspiration for solar power is present among groups across the political spectrum.

The many different proposals for a Green New Deal (GND) also reflect a common convergence around renewable energy technologies. The GND is an "ecology-centered economic stimulus program" with focus on renewable energy technology, economic growth, efficiency improvements, and creation of "green" jobs (Chohan, 2019). In the United States, the GND has primarily been driven by social democrats occupying a moderate position between proponents of capitalist growth and advocates of degrowth (Dale, 2019). Currently, however, it is favored both in neoliberal proposals for "green growth" and in eco-socialist visions of an "ecological politics for the working class" (Schwartzman and Schwartzman, 2013; Huber, 2019). From this, we can conclude either that the respective ideologies share a common philosophy or that there is a confusion regarding the politics and feasibility to transform the modern human-environmental condition through "green" technologies. While the former conclusion appears farfetched, there does indeed seem to be a confusion regarding the transformative potential of technology both in mainstream and critical theory (see Chapter 2).

Despite this confusion, the contradictions of renewable energy technologies have not gone unnoticed. A significant body of literature has emerged that critically examines the social and ecological conditions of different renewable energy technologies, including solar power, hydroelectric power, wind power, and biofuels. Here I attempt to summarize some of the most relevant findings of this rapidly growing literature:

- *Renewable energy technologies require comparatively large amounts of land per energy unit harnessed* (de Castro et al., 2013; Smil, 2015; Capellán-Pérez et al., 2017). There is strong evidence that solar power, wind power, hydroelectric power, and biofuels require large amounts of horizontal surface area compared to fossil fuels (Smil, 2015). Renewable energy sources such as sunshine, wind, and water are dispersed flows of energy that are not highly concentrated in subterranean deposits like coal, oil, or fossil gas. Since flows of energy are less concentrated, technologies capturing them must be dispersed across large surface areas.
- *Renewable energy technologies require comparatively large amounts of energy per energy unit harnessed* (Hall and Klitgaard, 2012; Hall et al., 2014; Ferroni and Hopkirk, 2016). All renewable energy technologies – with the exception of hydropower– require a large energy investment per unit of energy return. By extension, these studies show that it will be problematic to maintain advanced industrial societies solely with the help of renewable energy technologies.
- Large-scale manufacturing, transportation, and installation of renewable energy *technologies require significant amounts of fossil energy and imply greenhouse gas emissions* (Fu et al., 2015; Stamford and Azapagic, 2018). The new energy regime can only emerge from the energetic and material basis of the old one, i.e., by burning fossil fuels. This fact gives rise to the dilemma known as the "energy-emissions trap" that states that economies seeking to transition away from high net energy fossil fuels to lower net energy renewables for the sake of reducing greenhouse gas emissions must significantly increase the emissions of greenhouse gases (Sers and Victor, 2018).
- There are some studies suggesting that there are *limits to the quantity and quality of materials* needed for massively scaling up the installation of renewable energy technologies (de Castro et al., 2013; Tokimatsu et al., 2017; Rhodes, 2019). For solar power, this includes potential limits to recoverable reserves of silver, tellurium, indium, and selenium. Given the geographical dispersal of high-quality materials and the fact that "low-hanging fruits" are picked first, the finite nature of mineral deposits will likely drive increasing levels of geopolitical conflicts across the globe as renewable energy technologies are commercialized (Vakulchuk et al., 2020).
- *The installation of renewable energy technologies is not strictly speaking replacing fossil technologies* (York, 2012; York and Bell, 2019; Gellert and Ciccantell, 2020). Since the consumption of fossil energy carriers is increasing, installing more renewable energy technologies amounts to an energy addition, rather than an energy transition. As we will see later in this book, most of the ambitious plans to increase the relative level of solar power in the energy

mix simultaneously imply a total increase in fossil fuel use. Considering that modern iron and steel production necessitate coke (pyrolyzed coal), fossil fuels cannot be completely replaced with renewable energy (Smil, 2009, 2016).

- In the global commodity chains, there are *negative and unequally distributed health effects and environmental consequences of renewable energy technologies* (Dubey et al., 2013; Yue et al., 2014; Nain and Kumar, 2020). These effects include toxicity and environmental pollution in the manufacturing, transportation, and end-of-life processes. The effects disproportionally affect people who do not immediately benefit from the energy harnessed by the technologies. Similar injustices have been observed in state-backed enclosure of common land and dispossession of vulnerable communities to make way for utility-scale renewable energy parks producing energy for urban populations (e.g., Yenneti et al., 2016; Stock and Birkenholtz, 2019a, 2019b).
- In the process of mining and producing more metals for renewable energy technologies, there are *significant risks to plants and animals in local environments*. These "new threats to biodiversity," one study concluded, "may surpass those averted by climate change mitigation" (Sonter et al., 2020). Rehbein et al. (2020) found that solar power, in comparison to other renewable energy technologies, is more likely to be installed in protected areas even if solar irradiation is widely available in low-biodiversity lands.

This scientific literature shows that it is not at all clear that renewable energy technologies can contribute to the environmentally sustainable and socially just energy regime that so many envision. Thus, the benefits of solar power appear contradictory: On the one hand, solar power might embody the transformative potential to inaugurate new social-ecological relations by mitigating climate change, providing clean renewable energy, and encouraging democratic forms of ownership and decision-making. On the other hand, solar technologies may imply notable amounts of greenhouse gas emissions and other forms of pollution, depletion of nonrenewable resources, global wage gaps, and non-democratic conditions along the entire commodity chain. This multifaceted contradiction is significant because it forces us to question the conventional understanding of technology as "applied science" and what it means to choose renewable energy technologies in the hope of creating a better world.

Despite this, there exists an optimistic view based on an unshaken belief in technological progress that tends to deny, downplay, or even ridicule the observed downsides to solar power (e.g., Bastani, 2019; Phillips, 2020). In this view – commonly labeled *ecological modernism* – new technologies, efficiency improvements, material/resource substitutes, and engineering design are frequently invoked as means to counter socially articulated concerns over technological problems or contradictions. Ecomodernists think of the environmental degradation and socially eroding consequences of a particular technology as temporarily necessary (Levidow et al., 2012: 159). Technologies, in this view, are never inherently problematic, because they are continually improved, redesigned, or replaced by other technologies with different material components and social implications. In the discussions on the energy return of solar power, for example, the notion of technological progress is invoked to explain how solar PV cells can be made more energy efficient in the future and thereby fulfill one of the key criteria for feasibly sustaining advanced industrial societies (see Chapter 5). Ecological modernism offers a comfortable narrative rooted in a strong belief in technological progress suggesting that no significant social change is necessary for solving the ecological crisis. Any concerns regarding negative repercussions of "green" technology are brushed aside on the basis that human ingenuity represents an infinite creative and productive force (Goldstein, 2018). The alleged evidence for this infinite force is the rapid development that industrial nations have enjoyed since the Industrial Revolution (cf. Barca, 2011). This position is coupled with a teleological notion of history as a progressive journey of human mastery over nature, where an infinitely malleable nature exists to serve human ends.

For those who are unwilling to believe in technological progress, the contradictions associated with solar power represent significant risks to the creation of a socially just and environmentally sustainable world. These ecological realists (as we may call them) acknowledge contradictions and question the validity and desirability of technological – as opposed to social-political or cultural – solutions to environmental problems (e.g., Huesemann and Huesemann, 2011; Ruuska et al., 2020). It is true that the "consumption" of resources and environmental impacts have massively increased in the world economy *despite* two centuries of technological progress (Steffen et al., 2015; Wiedmann et al., 2015). Considering this historical evidence, ecological realists do not assume that technologies which have not been implemented in the world will necessarily solve environmental problems. Emphasis is placed on caution, risk prevention, and understanding rather than risk taking and a rush to deploy new technologies for the sake of technological progress itself (e.g., Gowdy, 1994). In this view, which Earl Cook (1976) once labeled the "prudence option," problems are unlikely to be solved with technological solutions that are dependent upon the same relations, the same forces, and the same type of thinking that generated them. Many ecological realists argue that there needs to be a societal-wide convergence around the ambition to create a way of living in which less is produced, less is consumed, and life is simpler, slower, more caring, and more communal (e.g., Trainer and Alexander, 2019). In relation to solar power, this position is exemplified in a study by Capellán-Pérez et al. (2020), which shows that the low energy return of solar PV technology means that solar power cannot support advanced industrial societies. Even if this is so, the authors argue, advanced industrial societies could in fact exist in the future, but only if solar PV modules are complemented with fossil fuels (see also Georgescu-Roegen, 1978). Thus, the contradictions of solar power serve to question the trajectory of advanced industrial societies based on the social condition to endlessly exploit the human and natural world.

A note on interdisciplinarity

In the process of writing, presenting, and discussing this book, I have realized that the reader deserves a clarification of the book's interdisciplinary approach. This realization came in part from colleagues and friends who told me so quite explicitly and in part from self-reflection on how to communicate the book's message effectively to readers who do not have an interdisciplinary background.

The first step in this clarification is to repeat that this book is a work within the interdisciplinary field of human ecology. Human ecology is a field of study that integrates insights from disciplines such as biology, physics, sociology, anthropology, economics, history, psychology, and philosophy in order to reach a holistic understanding of human-environmental relations (Steiner and Nauser, 1993: 2–6). This means that the reader should be prepared that concepts from any of these disciplines could be presented, employed, and critically discussed. Among these disciplines, many have "environmental" or "ecological" offshoots of their own, with relevance for understanding solar PV technology. In particular, I draw upon concepts and insights from ecological economics, philosophy of technology, environmental sociology, environmental history, geography, environmental anthropology, and to a lesser extent environmental psychology.

The second step is to clarify what an interdisciplinary approach is and what purpose it serves. One definition of interdisciplinarity is that it involves "bringing together distinctive components of two or more disciplines" (Nissani, 1997: 203).[5] A discipline is "any comparatively self-contained and isolated domain of humane experience which possesses its own community of experts" (ibid. 203). Interdisciplinarity is a process that includes the cognitive task of "integration," which involves "critically evaluating disciplinary insights and creating common ground among them" (Repko and Szostak, 2017: 669). Whereas the disciplinary approach studies a range of cases, events, or things in order to develop a particular concept describing the world, the interdisciplinary approach integrates concepts from a variety of disciplines to understand one particular case, event, or thing in the world (Figure 1.1). Emphasis is placed on general knowledge as opposed to specialized knowledge; integration as opposed to separation; and concept synthesis as opposed to concept analysis.

The purpose of the interdisciplinary approach is to generate a more comprehensive knowledge of the phenomena studied. As such, the interdisciplinary

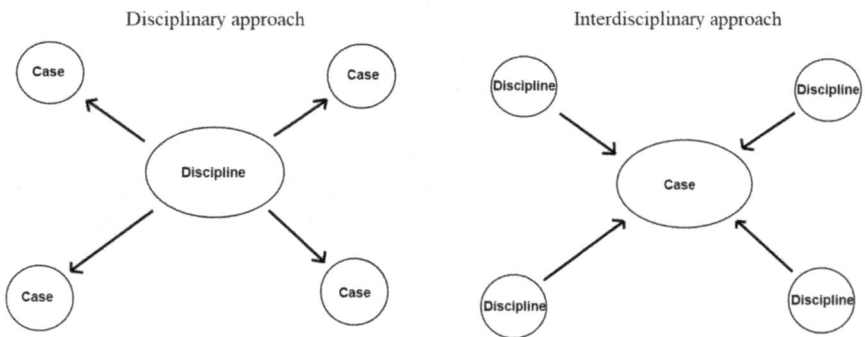

Figure 1.1 A basic illustration of the disciplinary and interdisciplinary approaches. The arrows represent knowledge and/or concept application.

approach encourages an epistemological transformation in the researcher during which numerous perspectives are eventually taken as a multitude of languages for expressing the same single phenomenon. Interdisciplinarity has the potential to generate a "hitherto non-existent connection" that cannot be generated from the perspectives of each of the disciplines it draws upon (Jahn et al., 2012: 9). The disciplinary parts are integrated into a whole that is qualitatively distinct from the separate parts taken together. This book seeks to expose precisely such "hitherto non-existent connections" of solar PV technology.

The third step is to clarify the strengths and weaknesses of the interdisciplinary approach. The perhaps most persuasive strength of interdisciplinarity is that reality – contrary to modern scientific attempts to understand it – is not compartmentalized into disciplines. "These [disciplinary] boxes," as Wallerstein (2004: x) called them, are in fact "constructs more of our imagination than of reality." Moreover, a compartmentalized view of reality encourages professionals who, as Nissani put it (1997: 209), know "everything about the chemistry of CFCs yet nothing about the ozone layer," or "everything about internal combustion engines and nothing about global warming." Rigid disciplinarity may thereby create a situation in which scientific experts become alienated from the wider effects of their expertize (Jahn et al., 2012). One strength of interdisciplinarity is that it counters this situation.

The perhaps most important weakness of interdisciplinarity is that it aspires to translate "the chaos of babel," as Steiner and Nauser (1993: 17) put it. This weakness, to be more concrete, is that each discipline comes with an often-unique terminology, which prevents clear and effective communication across disciplines (ibid. 17). Pellmar and Eisenberg (2000: 43) articulate this challenge by pointing out how "extensive effort must be made to learn the language [terminology] of another field and to teach others the language of one's own" before "successful collaboration can occur." If the writer-reader relation is anything like a collaboration, then this means that interdisciplinarity may demand both increased efforts to communicate (on the part of the writer) and to understand what is written (on the part of the reader). For this reason, I have included a list with explanations of the most frequently recurring concepts in the book (see the Glossary).

The structure and contents of this book

This book consists of seven chapters. In the next chapter, I clarify the methodological point of departure and the theoretical framework of the book. I first give a brief overview of philosophical materialism and the key relevance it has for understanding the biosphere and the current state of the planet. I then present my unanticipated discovery that the most prominent philosophies of technology have not been concerned with understanding technology as something contingent on matter-energy in the physical sense (ontological materialism). It follows that the interpretation of nature that is today employed to understand the dire state of the planet is broadly absent in the conception of solar PV technology as one of the most favored solutions. I propose that there is much to gain by allowing philosophical materialism to inform the understanding of technology. I attempt to show this by suggesting an interdisciplinary framework for understanding technology that is commensurable with the philosophical assumptions guiding research on the biosphere. The resultant "critical ecological philosophy of technology" invites us to consider modern technology as intertwined with the socially organized exchange of matter-energy that began with the Industrial Revolution. This notion is captured in a conceptual model that I call "the technological continuum," which I will frequently apply throughout the book.

In the third chapter, I contextualize visions of transforming society by means of solar power by considering how the technology emerges from the historical context of industrial capitalism propelled by fossil fuels. As argued by environmental historians, large-scale development of solar PV technology and other renewable energy technologies is conventionally envisioned to form the technological basis of an entirely new social-ecological regime. Such a transformation is analogous to historical events such as bipedalism, the adoption of fire, the Neolithic Revolution, and the Industrial Revolution. To understand what it means to transition away from the industrial regime based on fossil fuels, I place particular emphasis on accounting for the socio-ecological conditions of the Industrial Revolution from which modern technologies first arose. I highlight how it is difficult to separate modern technology from some of the defining features of the industrial regime (other than analytically). This, as I show, is only one of at least three interpretations in the discussion of the potential of solar PV technology to inaugurate a new energy regime. At stake is whether large-scale development of solar PV technology can be said to continue, transcend, or reverse the social-ecological conditions of the industrial regime.

The fourth chapter deals with the recent and rapid developments in the global solar PV industry. Between the years 2000 and 2018, solar PV technology developed from being an alternative technology to becoming a fully-fledged commercial commodity on the world market. I show that the most common explanations for why and how this boom occurred have largely omitted any meaningful consideration of material flows in international trade. Through an LCA-based assessment of ecologically unequal exchange, I examine the biophysical link between the two crucial actors Germany and China during the booming years. The results demonstrate that an ecologically unequal exchange – whereby demands on labor, land, energy, and greenhouse gas emissions could be displaced to China – was necessary for Germany to fulfill its solar visions at the time. This demonstrates that the recent low prices of solar PV modules are intimately connected to the contemporary world division of labor. From a broad perspective, this means that the notion of solar technology, as an "alternative technology" with promising democratic and sustainability potential, must be reconsidered based on its potentially inherent global asymmetries.

In Chapter 5, I discuss the question of whether and in what sense global asymmetric transfers of resources are inherent to large-scale solar PV technology projects. I discuss this in relation to Langdon Winner's (1980) idea of the politics of technological artifacts. Drawing on methodological discussions on how to measure "energy return on energy investment" and "power density," I raise the question of how drawing boundaries around what constitutes solar technology can affect whether we perceive large-scale solar PV development as inherently political or not. By calculating the "power density extended" of five national solar aspirations (China, the US, Germany, India, and Italy), I demonstrate that a technological boundary that extends further back in the global commodity chain shows that substantial amounts of indirect land requirements are necessary for fulfilling modern solar PV aspirations. In this way, the large-scale solar PV projects of China, the US, Germany, India, and Italy all necessitate a highly politicized world division of labor as much as they necessitate polysilicon, engineers, electrical components, and direct sunshine. In this view, the world division of labor is an integral part of a new metabolism based on large-scale construction of solar PV parks. In contrast, a narrow technological boundary that perceives solar PV modules merely as artifacts independent from society cannot see this political condition as inherent in solar PV technology. I argue that the low "energy return on energy investment" and "power density" of solar PV technologies should not be understood as something that renders a large-scale transition to solar PV technology impossible, but that it may require (not merely encourage) unequal distribution of social and environmental burdens and benefits in the global economy that are problematic for the ongoing transition away from fossil fuels. Importantly, these problems will not be visible to us as long as we retain a narrow understanding of what solar PV technology, and indeed technology, *is*.

In the sixth chapter, I reflect on the questions raised in this introductory chapter and discuss the results in relation to the theoretical point of departure. The results of the book suggests that ecologically unequal exchange is likely a prerequisite for practically realizing conventional solar visions and how this condition is repeatedly overlooked or denied by researchers, policymakers, corporations, and governments who are working toward a low-carbon transition. Thus, solar PV technology is being embraced without fully considering its global prerequisites. I suggest that this is a symptom of fetishization, whereby the biophysical history of solar PV modules is rendered immaterial and "forgotten," and how alienation, power, ideology, and denial may be the determinants for this forgetting. I raise the question whether the strategic omission of the global distributive dimension of solar PV technology could one day be compared to ExxonMobil's strategic denial of climate change. Finally, through an organic analogy, I discuss the ontology of technology and provide a definition of technology that can be employed to avoid this problem.

In the seventh and final chapter, I argue that the results of the previous chapters may create a sense of disillusionment from conventional solar visions and how this sense is best countered with alternative action-inspiring solutions. Such solutions, I argue, may even be necessary for people to be able to acknowledge machine fetishism and to understand the conclusions of this book. As long recognized by environmentalists, to face the reality of the ecological crisis is to live in a "world of wounds," which explains the appeal to hide from threatening information by resorting to unrealistic visions and machine fetishism. To break from machine fetishism, I introduce the notion of "realistic envisioning" as a means for local communities, grassroots organizations, and policymakers to discover the energy technologies suitable for desirable, just and sustainable futures. I then discuss the social-ecological ramifications of an "alternative solar technology" based on photosynthesizing plants as a theme for realistic envisioning. I conclude with a call for alliances across subsistence- and resistance-oriented movements to explore the potential of a "metabolic counter-regime" antithetical to ecologically unequal exchange and its associated global environmental injustices.

Notes

1   The aim is not to compare whether the ecological effects of fossil fuels present a better or worse ecological situation than the effects of solar power. It is abundantly clear that fossil fuels are finite and non-sustainable stocks of energy that the world's actors must immediately stop burning in order to halt the detrimental effects of climate change (IPCC, 2014, 2018).
2   So-called "greenwashing" – whereby corporations falsely brand commodities as environmentally sustainable to increase the rate of profit – is a well-known phenomenon by which misleading symbolism obscures the real-world impacts of commodities (see, e.g., Lyon and Montgomery, 2015).
3   For a description of the term "embodied," see the Glossary.
4   This name is a pseudonym.
5   Transdisciplinarity, contrary to interdisciplinarity, includes social groups outside academia as part of the research process (Jahn et al., 2012). By including social groups with knowledge that cannot be reduced to disciplinary categorization, transdisciplinarity transcends the very systematization upon which scientific disciplines are based.

References

Asafu-Adjaye, John, Linus Blomqvist, Stewart Brand, Barry Brook, Ruth DeFries, Erle Ellis, Christopher Foreman, et al. 2015. *An ecomodernist manifesto.* www.ecomodernism. org.

Barca, Stefania. 2011. Energy, property, and the industrial revolution narrative. *Ecological Economics* 70: 1309–1315. https://doi.org/10.1016/j.ecolecon.2010.03.012.

Bastani, Aaron. 2019. *Fully automated luxury communism.* London, England and New York, NY: Verso.

Bonds, Eric, and Liam Downey. 2012. "Green" technology and ecologically unequal exchange: The environmental and social consequences of ecological modernization in the world-system. *American Sociological Association* 18(2): 167–186. https://doi. org/10.5195/jwsr.2012.482.

Buchan, David. 2012. *The Energiewende—Germany's gamble.* Oxford: Oxford Institute for Energy Studies.

Capellán-Pérez, Iñigo, Carlos de Castro, and Iñaki Arto. 2017. Assessing vulnerabilities and limits in the transition to renewable energies: Land requirements under 100% solar energy scenarios. *Renewable and sustainable energy reviews* 77: 760-782. https://doi.org/ 10.1016/j.rser.2017.03.137.

Chohan, Usman W. 2019. A green new deal: Discursive review and appraisal. *Notes on the 21st Century* (CBRI). https://ssrn.com/abstract=3347494.

Cook, Earl F. 1976. *Man, energy, society*. Texas: Texas A. and M. University, College of Geosciences, College Station.

Dale, Gareth. 2019. Degrowth and the Green New Deal. *The Ecologist*. Published 28-10-2019. https://theecologist.org/2019/oct/28/degrowth-and-green-new-deal. Accessed 05-22-2020.

de Castro, Carlos, Margarita Mediavilla, Luis Javier Miguel, and Fernando Frechoso. 2013. Global solar electric potential: A review of their technical and sustainable limits. *Renewable and Sustainable Energy Reviews* 28: 824–835. https://doi.org/10.1016/j.rser.2013.08.040.

Diaz-Maurin, François, and Mario Giampietro. 2013. A "grammar" for assessing the performance of power-supply systems: Comparing nuclear energy to fossil energy. *Energy* 48: 162–177. https://doi.org/10.1016/j.energy.2012.11.014.

Dubey, Swapnil, Nilesh Y. Jadhav, and Betka Zakirova. 2013. Socio-economic and environmental impacts of silicon based photovoltaic (PV) technologies. *Energy Procedia* 33: 322–334. https://doi.org/10.1016/j.egypro.2013.05.073.

Ellul, Jacques. 1964. *The technological society*. New York, NY: Vintage.

Feenberg, Andrew. 1991. *Critical theory of technology*. New York, NY: Oxford University Press.

Ferroni, Ferruccio, and Robert J. Hopkirk. 2016. Energy return on energy invested (ERoEI) for photovoltaic solar systems in regions of moderate insolation. *Energy Policy* 94: 336–344. https://doi.org/10.1016/j.enpol.2016.03.034.

Fu, Yinyin, Xin Liu, and Zengwei Yuan. 2015. Life-cycle assessment of multi-crystalline photovoltaic (PV) systems in China. *Journal of Cleaner Production* 86: 180–190. https://doi.org/10.1016/j.jclepro.2014.07.057.

Gellert, Paul K., and Paul S. Ciccantell. 2020. Coal's persistence in the capitalist world-economy: Against teleology in energy 'transition' narratives. *Sociology of Development* 6(2): 194–221. https://doi.org/10.1525/sod.2020.6.2.194.

Georgescu-Roegen, Nicholas. 1978. Technology assessment: The case of the direct use of solar energy. *Atlantic Economic Journal* 6: 15–21. https://doi.org/10.1007/BF02300267.

Giampietro, Mario, and Kozo Mayumi. 2015. *The biofuel delusion: The fallacy of large scale agro-biofuels production*. New York, NY: Routledge.

Goldstein, Jesse. 2018. *Planetary improvement: Cleantech entrepreneurship and the contradictions of green capitalism*. Cambridge and London: The MIT Press.

Gowdy, John M. 1994. Progress and environmental sustainability. *Environmental Ethics* 16: 41–55. https://doi.org/10.5840/enviroethics199416140.

Hall, Charles A. S., and Kent A. Klitgaard. 2012. *Energy and the wealth of nations: Understanding the biophysical economy*. New York, NY: Springer.

Hall, Charles A. S., Jessica G. Lambert, and Stephen B. Balogh. 2014. EROI of different fuels and the implications for society. *Energy Policy* 64: 141–152. https://doi.org/10.1016/j.enpol.2013.05.049.

Heidegger, Martin. 1977[1954]. *The Question Concerning Technology and Other Essays*. New York, NY and London, England: Garland Publishing.

Hornborg, Alf. 2006. Footprints in the cotton fields: The Industrial Revolution as time-space appropriation and environmental load displacement. *Ecological Economics* 59(1): 74–81. https://doi.org/10.1016/j.ecolecon.2005.10.009.

Hornborg, Alf, Gustav Cederlöf, and Andreas Roos. 2019. Has Cuba exposed the myth of "free" solar energy? Energy, space, and justice. *Environment and Planning E: Nature and Space* 2(4): 989–1008. https://doi.org/10.1177/2514848619863607.

Huber, Matthew T. 2019. Ecological politics for the working class. *Catalyst* 3(1).

Huesemann, Michael, and Joyce Huesemann. 2011. *Techno-fix: Why technology won't save us or the environment*. Gabriola, BC, Canada: New Society.

Illich, Ivan. 1973. *Tools for conviviality*. Glasgow, Scotland: Collins.

IPCC. 2014. *Climate Change 2014: Synthesis report*. Contribution of working groups I, II and III to the fifth assessment report of the intergovernmental panel on climate change. IPCC, Geneva, Switzerland.

IPCC. 2018. Summary for policymakers. In *Global warming of 1.5°C. An IPCC special report on the impacts of global warming of 1.5°C above pre-industrial levels and related global greenhouse gas emission pathways, in the context of strengthening the global response to the threat of climate change, sustainable development and efforts to eradicate poverty*, edited by Masson-Delmotte, V., P Zhai, H.O. Pörtner, D. Roberts, J. Skea, P.R. Shukla, A. Pirani, et al. 3-24. Cambridge and New York: Cambridge University Press. https://doi.org/10.1017/9781009157940.

Jahn, Thomas, Matthias Bergmann, and Florian Keil. 2012. Transdisciplinarity: Between mainstreaming and margnialization. *Ecological Economics* 39: 1–10. https://doi.org/10.1016/j.ecolecon.2012.04.017.

Jeffries, Elisabeth. 2017. Nationalist advance. *Nature Climate Change* 7: 469–471. https://doi.org/10.1038/nclimate3334.

Klein, Naomi. 2014. *This changes everything: Capitalism vs. the climate*. New York, NY: Penguin Random House Knopf Canada.

Levidow, Les, Theo Papaioannaou, and Kean Birch. 2012. Neoliberalizing technoscience and environment: EU policy for competitive, sustainable biofuels. In *Neoliberalism and technoscience: Critical assessments*, edited by Luigi Pellizzoni and Marja Ylönen, 159–186. New York, NY: Routledge.

Lyon, Thomas P., and A. Wren Montgomery. 2015. The means and ends of greenwash. *Organization and Environment* 28(2): 223–249. https://doi.org/10.1177/1086026615575332.

Malm, Andreas. 2016. *Fossil capital: The rise of steam power and the roots of global warming*. London, England: Verso.

Mander, Jerry. 1991. *In the absence of the sacred: The failure of technology and the survival of the Indian nations*. San Francisco, CA: Sierra Club.

Marx, Leo. 2010. Technology: The emergence of a hazardous concept. *Technology and Culture* 51(3): 561–577.

Miller Tyler G., and Scott E. Spoolman. 2009. *Living in the environment: Concepts, connections and solutions*. 16th ed. Belmont, CA: Brooks/Cole.

Mulvaney, Dustin. 2019. *Solar power: Innovation, sustainability and environmental justice*. Oakland: University of California Press.

Nain, Preeti, and Arun Kumar. 2020. Ecological and human health risk assessment of metals leached from end-of-life solar photovoltaics. *Environmental Pollution* 267: 115393. https://doi.org/10.1016/j.envpol.2020.115393.

Nemet, Gregory. 2019. *How solar energy became cheap: A model for low-carbon innovation*. London, England and New York, NY: Routledge.

Nissani, Moti. 1997. Ten cheers for interdisciplinarity: The case for interdisciplinary knowledge and research. *The Social Science Journal* 34(2): 201–216. https://doi.org/10.1016/S0362-3319(97)90051-3.

Pellmar, Terry C., and Leon Eisenberg, eds. 2000. *Bridging disciplines in the brain, behavioural, and clinical sciences*. Washington, DC: National Academy Press.

Phillips, Leigh. 2020. Planet of the anti-humanists. *Jacobin*. Published 05-04-2020. https://jacobinmag.com/2020/05/planet-of-the-humans-michael-moore-documentary-climate-change/. Accessed 06-11-2020.

Rehbein, Jose A., James E. M. Watson, Joe L. Lane, Laura J. Sonter, Oscar Venter, Scott C. Atkinson, and James R. Allan. 2020. Renewable energy development threatens many globally important biodiversity areas. *Global Change Biology* 26(5): 3040–3051. https://doi.org/10.1111/gcb.15067.

Repko, Allen F., and Rick Szostak. 2017. *Interdisciplinary research: Process and theory.* 3rd ed. Los Angeles, CA: Sage.

Rhodes, Christopher J. 2019. Endangered elements, critical raw materials and conflict minerals. *Science Progress* 102(4): 304–350. https://doi.org/10.1177/0036850419884873.

Ruuska, Toni, Pasi Heikkurinen, and Kristoffer Wilén. 2020. Domination, power, supremacy: Confronting anthropolitics with ecological realism. *Sustainability* 12(7): 2617. https://doi.org/10.3390/su12072617.

Scheer, Hermann. 2007. *Energy autonomy: The economic, social and technological case for renewable energy.* London, England and Sterling, VA: Earthscan.

Schwartzman, David. 1996. Solar communism. *Science and Society* 60(3): 307–331.

Schwartzman, David, and Peter Schwartzman. 2013. A rapid solar transition is not only possible, it is imperative! *African Journal of Science, Technology, Innovation and Development* 5(4): 297–302.

Sers, Martin R., and Peter A. Victor. 2018. The energy-emissions trap. *Ecological Economics* 151: 10–21. https://doi.org/10.1016/j.ecolecon.2018.04.004.

Shepard, Paul. 1967. Whatever happened to human ecology? *Bioscience* 17(12): 891–911. https://doi.org/10.2307/1293928.

Smil, Vaclav. 2009. The Iron Age and coal-based coke: A neglected case of fossil-fuel dependence. *Master resource.* Published 17-09-2009. https://www.masterresource.org/smil-vaclav/a-forgotten-case-of-fossil-fuel-dependence-the-iron-age-requires-carbon-based-energy-like-it-or-not/. Accessed 28-12-2020.

Smil, Vaclav. 2015. *Power density: A key to understanding energy sources and uses.* Cambridge, MA: MIT Press.

Smil, Vaclav. 2016. *Still the Iron Age: Iron and steel in the modern world.* Oxford, England and Cambridge, MA: Butterworth-Heinemann.

Sonter, Laura J., Marie C. Dade, James E. M. Watson, and Rick K. Valenta. 2020. Renewable energy production will exacerbate mining threats to biodiversity. *Nature Communications* 11: 4174. https://doi.org/10.1038/s41467-020-17928-5.

Sovacool, Benjamin K. 2011. *Contesting the future of nuclear power: A critical global assessment of atomic energy.* Hackensack, NJ and London, England: World Scientific.

Stamford, Laurence, and Adisa Azapagic. 2018. Environmental impacts of photovoltaics: The effects of technological improvements and transfer of manufacturing from Europe to China. *Energy Technology* 6: 1146–1160. https://doi.org/10.1002/ente.201800037.

Steffen, Will, Wendy Broadgate, Lisa Deutsch, Owen Gaffney, and Cornelia Ludwig. 2015. The trajectory of the Anthropocene: The great acceleration. *The Anthropocene Review* 2(1): 81–98. https://doi.org/10.1177/2053019614564785.

Steiner, Dieter, and Markus Nauser, eds. 1993. *Human ecology: Fragments of anti-fragmentary views of the world.* London, England and New York, NY: Routledge.

Stock, Ryan, and Trevor Birkenholtz. 2019a. The sun and the scythe: Energy dispossessions and the agrarian question of labor in solar parks. *The Journal of Peasant Studies* 48(5): 1–24. https://doi.org/10.1080/03066150.2019.1683002.

Stock, Ryan, and Trevor Birkenholtz. 2019b. Photons vs. firewood: Female (dis)empowerment by solar power in India. *Gender, Place and Culture* 27(1): 1628–1651. https://doi.org/10.1080/0966369X.2020.1811208.

Tokimatsu, Koji, Henrik Wachtmeister, Benjamin McLellan, Simon Davidsson, Shinsuke Murakami, Mikael Höök, Rieko Yasuoka, and Masahiro Nishio. 2017. Energy modeling approach to the global energy-mineral nexus: A first look at metal requirements and the 2°C target. *Applied Energy* 207(1): 494–509. https://doi.org/10.1016/j.apenergy.2017.05.151.

Trainer, Ted. 2014. Some inconvenient theses. *Energy Policy* 64: 168–174. https://doi.org/10.1016/j.enpol.2013.06.008.

Trainer, Ted, and Samuel Alexander. 2019. The simpler way: Envisioning a sustainable society in an age of limits. *Real-World Economics Review* 87: 247–260.

Vakulchuk, Roman, Indra Overland, and Daniel Scholten. 2020. Renewable energy and geopolitics: A review. *Renewable and Sustainable Energy Reviews* 122: 109547. https://doi.org/10.1016/j.rser.2019.109547.

Wallerstein, Immanuel. 2004. *World-systems analysis: An introduction*. Durham, NC and London, England: Duke University Press.

Wealer, Ben, Simon Bauer, Leonard Göke, Christian R. von Hirschhausen, and Claudia Kemfert. 2019. High-priced and dangerous: Nuclear power is not an option for the climate-friendly energy mix. *DIW Weekly Report* 9(30): 253–243.

Wiedmann, Thomas O., Heinz Schandl, Manfred Lenzen, Daniel Moran, Sangwon Suh, James West, and Keiichiro Kanemoto. 2015. The material footprint of nations. *PNAS* 112(20): 6271–6276. https://doi.org/10.1073/pnas.1220362110.

Winner, Langdon. 1978. *Autonomous technology: Technics-out-of-control as a theme in political thought*. Cambridge, MA: The MIT Press.

Winner, Langdon. 1980. Do artifacts have politics? *Daedalus* 109(1): 121–136. https://www.jstor.org/stable/20024652.

Winner, Langdon. 2020[1986]. *The whale and the reactor: A search for limits in an age of high technology*. Chicago, IL and London, England: The University of Chicago Press.

Yenneti, Komali, Rosie Day, and Oleg Golubchikov. 2016. Spatial justice and the land politics of renewables: Dispossessing vulnerable communities through solar energy megaprojects. *Geoforum* 76: 90–99. https://doi.org/10.1016/j.geoforum.2016.09.004.

York, Richard. 2012. Do alternative energy sources displace fossil fuels? *Nature Climate Change* 2(6): 441–443. https://doi.org/10.1038/nclimate1451.

York, Richard, and Shannon Elizabeth Bell. 2019. Energy transitions or additions? Why a transition from fossil fuels requires more than the growth of renewable energy. *Energy Research and Social Science* 51: 40–43. https://doi.org/10.1016/j.erss.2019.01.008.

Yue, Dajun, Fengqi You, and Seth B. Darling. 2014. Domestic and overseas manufacturing scenarios of silicon-based photovoltaics: Life cycle energy and environmental comparative analysis. *Solar Energy* 105: 669–678. https://doi.org/10.1016/j.solener.2014.04.008.

Zehner, Ozzie. 2012. *Green illusions: The dirty secret of clean energy and the future of environmentalism*. Lincoln and London, England: University of Nebraska Press.

2 Earthing philosophy of technology[1]

In the case of Katrineholm, we saw how solar developers have a hard time reconciling that access to solar photovoltaic (PV) modules may depend on an exploitation of cheap labor and lands elsewhere in the world economy. How do we study whether this is an inherent or transitory condition of solar PV technology? In this chapter, I argue that answering this question requires that we have a notion of reality and how it is filtered through cultural categories. This requires that we take a view of technology that is rooted in a recognition of nature as a physical phenomenon. The assertions of philosophical materialism have been pivotal in modern history for recognizing the existence of a self-orchestrating nature, which the ancient Greeks called *physis* and the Chinese called *ziran*. To this day, however, the same assertions have not been accepted or emphasized in the understanding of technology. This makes our inquiry difficult because it presents us with a situation wherein the understanding of nature and the understanding of technology are based on fundamentally different assumptions about the world. To remedy this situation, I side with the philosophical assertions that now guide research on the biosphere. I thereby situate the phenomenon of solar PV technology *within* a biophysical reality that is constituted by self-orchestrated processes of matter-energy. This point of departure makes it possible to study how ecologically unequal exchange – understood as an uneven transfer of resources masked by market valuation – relates to solar PV technology. This method can then function as a means to assess the inherent or transitory character of environmental injustices in the PV commodity chain, which will inform us about the validity of conventional solar visions.

The argument of this chapter is developed in three parts. I first give a brief overview of philosophical materialism and the key relevance it has for understanding the biosphere and the current state of the planet. I then show that the most prominent philosophies of technology over the last two centuries have not primarily been concerned with understanding technology as something contingent upon matter-energy in the physical sense (ontological materialism). It follows that the interpretation of nature that is employed to understand the state of the biosphere is broadly absent in the articulation of the technologies meant to protect and improve it, such as solar PV technology. Third, I will propose that philosophy of technology, and indeed humanity, has much to gain by allowing

DOI: 10.4324/9781003292319-2

philosophical materialism to inform the conception of technology. I attempt to show this by suggesting a philosophical framework for understanding technology that is commensurable with the philosophical assumptions underscoring research on the biosphere. I synthesize this "critical ecological philosophy of technology" primarily from the works of Lewis Mumford, Alf Hornborg, and John Bellamy Foster. It invites us to consider solar technology as intertwined with the human-environmental exchange of matter-energy that started with the Industrial Revolution and greatly accelerated with the access to oil. From this methodological and philosophical framework, I develop a pre-analytic model of technology that I call "the technological continuum" that will follow the reader throughout the book.

Does nature matter?

A general commitment to philosophical materialism implies an acknowledgment that all processes and things arise as part of the material world. All that is, in short, is of this world as it emerges from physical reality. In this view, even human thought is a physical expression of nature's internal processes. As poetically pointed out by Loren Eiseley (1969: 52), even "the human mind burns by the power of a leaf." Materialism in the philosophical sense must therefore be distinguished from the more common understanding of materialism as a collection of values for accumulating desirable artifacts ("to be materialistic"). Following critical realist Roy Bhaskar (1991: 369), philosophical materialism consists of three foundational statements:

a   *Ontological materialism*, asserting the unilateral dependence of social upon biological (and more generally physical) being and the emergence of the former from the latter;
b   *Epistemological materialism*, asserting the independent existence and trans-factual activity of at least some of the objects of scientific thought;
c   *Practical materialism*, asserting the constitutive role of human transformative agency in the reproduction and transformation of social forms.

Materialism in the philosophical sense has been lively debated for at least 2,000 years and is typically contrasted with "idealism," asserting that what exists are ideas. An important aspect of the material position is its formal opposition to religious explanations of nature and society that ascribe tangible powers to entities without substance in the world. While the rise of the material-mechanistic understanding of nature in Renaissance Europe has been identified as a central historical event underscoring modern ecological problems (Merchant, 1980), paradoxically, it is in the rejection of God that materialism bears a positive significance for an ecological understanding of the world. As John Bellamy Foster (2000) has demonstrated, philosophical materialism catalyzed ecological ways of thinking about the world. The tenets that today are central for ecology, Foster shows, can be directly related to the teachings of the ancient Greek materialist

Epicurus. These include that (i) everything is connected to everything else, (ii) everything must go somewhere, (iii) nature (or evolution) knows best, and (iv) nothing comes from nothing (ibid. 14). The full historical ebb and flow of these material-ecological tenets will not be reviewed here. Suffice to say, the historical prominence of materialism has fluctuated (for an overview, see Lange, 1925).

One important breakthrough for philosophical materialism came during the 19th century. This breakthrough is today mostly associated with the work of Charles Darwin (2011[1859]), who in his theory of evolution by natural selection contributed to an understanding of the variation of species because of natural-historical processes. The philosophical pathway for this theory had been cleared by earlier materialist thinkers both the organic and mechanical kind, such as De La Mettrie, Diderot, Holbach, and the Comte de Buffon. Even so, Darwin's materialism was highly controversial in a culture that for a long time had been dominated by theological explanations of the world (Clark et al., 2007). Even early materialists such as Isaac Newton, Francis Bacon, and Thomas Hobbes combined materialism with an acknowledgment of God (natural theology) that retained the understanding of nature as a static, non-changeable, phenomenon. In modern history, according to Engels (1964[1925]), it was first with Immanuel Kant's notion of nature as something that is "coming into being" and "passing away" that the mechanistic view of nature could be fundamentally challenged.[2] This discovery, Engels lauded, "contained the point of departure for all further progress" for an understanding of the world as historical-material (ibid. 27).

While Charles Darwin contribution to evolutionary biology might be the most famous example today, the insight that nature was undergoing internal processes of change was first taken up in the field of geology. Already in the early 1800s, Abraham Gottlob Werner had revolutionized the field of geology by demonstrating how differences in rocks existed due to differences in the time and mode of formation (for a discussion, see Foster, 2000: 117–122). In short, Werner's argument implied that different rock features did not simply exist statically in the world as orchestrated by a divine will. Rather, the Earth itself had formed them through geological processes spanning over long periods of time. One of the principal contributions of Werner was that it allowed for an understanding of the Earth as a historical entity characterized by self-organized changes not ordained by God. Reference to nature's internal processes, quite independent of any God, thus started to make sense as a basis for understanding an increasing range of phenomena in the world.[3]

At this time, a physicist by the name of John Tyndall took an interest in the peculiar geological evidence indicating that there had been a prehistoric Ice Age. Armed with a material epistemology, he suggested that the cause of the disappearance of this Ice Age might have been climatic changes. Tyndall supported his claim by demonstrating that water vapor had the capacity to prevent infrared radiation from dissipating into space and thereby creating a heating effect in the atmosphere. Consequently, he proposed that fluctuating levels of H_2O in the atmosphere could have been the cause for the observed climatic changes. Svante Arrhenius, a Swedish scientist, later showed that concentration of CO_2 in the

Earth's atmosphere, fluctuating in self-reinforcing feedback mechanisms, was a better explanation for climatic changes. As the physicist Spencer Weart (2003: 6) explained in *The Discovery of Global Warming*, even if Arrhenius was "far from proving how the climate would change if CO_2 varied, he did in truth get a rough idea of how it could change."

Today, the notion that nature is a material-historical process may seem obvious, even trivial. However, it is important to appreciate how the 19th-century materialist turn constituted a fundamental shift away from the earlier European understanding of nature as a static phenomenon created and organized by God. A new epistemology took hold that acknowledged nature itself as God-independent processes of great complexity. The very notion of the biosphere as a complex living system of biogeochemical cycles was later to be founded upon such a material understanding of the world (Suess, 1875; Vernadsky, 1998[1929]; Steffen et al., 2011). As the Soviet scientist Vladimir Vernadsky explained in *The Biosphere*:

> Creatures on Earth are the fruit of extended, complex processes, and are an essential part of a harmonious cosmic mechanism, in which it is known that fixed laws apply and chance does not exist. We arrive at this conclusion via our understanding of the matter of the biosphere (1998[1929]: 44).

Vernadsky's seminal work and understanding of the biosphere as a cosmic process with complex internal interactions between living and nonliving matter later influenced the development of Western ecology through G.E. Hutchinson, Eugene Odum, and Arthur Tansley in the 20th century (Oldfield and Shaw, 2013). What is more, Vernadsky contended, "all organisms are connected [to the biosphere] indissolubly and uninterruptedly, first of all through nutrition and respiration, with the circumambient material and energetic medium" (Vernadsky, 1945: 4). Importantly, humans too were regarded as organisms that were "indissolubly and uninterruptedly" part of the biosphere. In conversations with French scientists, Vernadsky proceeded to hypothesize that the biosphere was undergoing significant changes due to human influences. In the words of Steffen et al. (2011: 843), these "prophetic observers," therefore laid the foundation not only for the understanding of the biosphere, but even anticipated today's critical historical impasse. Contemporary deliberations on the "Anthropocene," "Technocene," or "Capitalocene" can all be interpreted as late variations to Vernadsky's observation that humans, "through [their] labor and [their] consciousness," are shifting the Earth into a new geological state (Vernadsky, 2001[1938]: 22).

This historical discussion simply scratches the surface of materialism and its significance for a recognition of nature.[4] It nevertheless demonstrates that the understanding that is today driving an increasing concern for the natural world is underlined by philosophical materialism. Materialism is in this sense an acknowledgment that nature matters in two interconnected ways:

1   Nature matters in the physical sense of constituting biogeochemical processes that exchange and transform matter and energy.

2 Nature matters in the sense of being necessary for the continuation of humans and other forms of life.

While it is important to reflect upon nature as a mental concept, it is essential that we acknowledge and recognize nature in the material sense (Soper, 1995; Ruuska et al., 2020). This is especially important for the aim to understand the contrast between social aspirations and ecological reality. In relation to the second point, human populations, like any other animal population, are part of the natural world, which they are dependent upon for their continuation (Daly, 1996; Heikkurinen et al., 2016). No organism can exist without an environment from which to draw matter-energy.

Today, the 19th-century materialist turn is important not simply because it reminds us that a fundamental shift in European epistemology occurred, but because the shift was never fully completed. For the purpose of this book, it is of particular relevance that this material turn was never fully applied to the understanding of technology. Even Vernadsky (1945: 4), who insisted that humans are "indissolubly" part of the biosphere, thought that technological progress was "the result of 'cephalization,' the growth of man's brain and the work directed by his brain." Karl Marx, who drew on the work of Charles Darwin, came closest to a materialist understanding of technology by speculating how it might be the human equivalent to "the formation of the organs of plants and animals, which serve as the instruments of production for sustaining their life" (Marx, 1990[1867]: 493, footnote 4). In the same footnote he wrote:

> Technology reveals the active relation of man to nature, the direct process of the production of his life, and thereby it also lays bare the process of the production of the social relations of his life, and of the mental conceptions that flow from those relations.

However, in his extensive corpus, Marx discusses this historical materialism of technology in only a single footnote. To the extent that we can draw a distinction between technology and machines, Marx was much more interested in machines than in technology.[5] For Marx, as for many of his contemporaries, the theme of the machine as a labor-saving device was of utmost interest. When reading Marx's most famous passages on the machine it is evident that Marx had an inclination to consider the productive potential of machines as originating from within the machines themselves (see, e.g., Marx, 1990[1867]: 494, 502; Marx, 1993[1939]: 818–819). In particular, machines regularly appear in Marx as if they have been simply conjured into existence with a ready-made productive potential. This is an understanding of technology that resembles the pre-materialist geology, where rock formations are conceived of as independent from their geological history.

Still today, I argue, the non-materialist view haunts contemporary analyses of technology. In effect, as I show further on in this chapter, modern conceptions of technology systematically miscalculate the feasibility to employ technology in the hope of solving environmental problems. This has subtle but important implications for understanding the feasibility of contemporary visions of future societies based on solar power. The rejection of materialist ontologies of technology is most obvious in philosophical frameworks where technology is explicitly understood as a "cognitive activity" or as a "consciousness." However, ontological materialism appears to be absent even in supposed materialist philosophies of technology. The following section is meant to demonstrate this surprising situation.

What is technology?

Let us now turn to contemporary philosophy of technology and ask what technology is. Among philosophers of technology, there have been countless ways of approaching this question (see Scharff and Dusek, 2014). Here, I will follow Andrew Feenberg's (1991, 2008) categorization, which distinguishes between three overarching theoretical approaches to technology. These are "instrumentalism," "substantivism," and "critical theory." To this list, I will add "actor-network theory." While this is not an exhaustive account of all existing philosophies of technology, it does provide a point of departure from which it is possible to engage with the 20th- and 21st-century discussions of technology.

Instrumentalism: a gift from the other

Instrumental theory is commonly pointed to as the dominant understanding of technology today (Feenberg, 1991: 5–7, 2008; Dusek, 2006: 53–69). This theoretical lens is employed for the most parts by governments and policymakers but is also common within the social sciences. At the core of scientific instrumentalism lie an anti-realist argument asserting that no theoretical explanation or concept can be claimed as explaining reality (Stanford, 2006). From this perspective, it is better to understand "theories as tools for pursuing practical ends" rather than as true descriptions of reality (ibid. 403).

In instrumental takes on technology, this practical imperative extends from theories to technologies, which must only be judged by the degree of efficiency with which they can be employed to solve problems. Concerns regarding the conditions and effects of technologies are typically brushed to the side as an external concern. As Dewey (2008: 354–355) put it:

> There is no problem of why and how the plow fits, or applies to, the garden, or the watch-spring to time-keeping. They were made for those respective purposes; the question is how well they do their work, and how they can be reshaped to do it better.

This position gives rise to the common, yet widely criticized notion that technologies are value-neutral (Winner, 1980; Huesemann and Huesemann, 2011: 235–241). Technological neutrality, or value-neutrality, is recognized as a position that can be captured in the cliché "guns don't kill people, people kill people."

Technologies, in this reductivist view, are independent from the contexts in which they arise and which they reproduce. Such a value-neutrality underscores Marx's famous critique of John Stuart Mill in the opening of Chapter 15 in *Capital* vol. 1 where Marx argues that the role of machinery under capitalism is not to save labor, but that machinery can take on this social role in other social modes.

The anti-realism that underscores instrumental theories of technology is key to understanding in what way technology is value-neutral. In Dewey's work, summed up by Larry Hickman (2014: 408, 410), technology is considered a "cognitive activity" that is "brought to bear on raw materials and intermediate stock parts, with a view to the resolution of perceived problems." This implies that technology is essentially an immaterial (cognitive) phenomenon, yet with tangible consequences in the material world.[6] This process is made possible through the technical operation of engineers, or "engineering design" (Mitcham, 1994: 225–228). Technology in the engineering sense is a result of the human ability to come up with different designs. "Designing," Mitcham (1994: 220) writes, can be identified as a process of extracting thoughts from the head of the engineers and delivering them into the real physical world via drawings, modelings, and blueprints (see also Layton, 1974: 38). Technology is consequently thought to originate from scientific knowledge and therefore understood as a form of applied science (Kline, 1995).

The notion that technology *is* design is similar to the European pre-materialist view of nature, in which nature's geological features were explained with reference to God's design. At times, the similarity is striking. For example, Frederich Dessauer, one of the founders of modern philosophy of technology, claimed that the mind of the engineer or inventor may be in contact with a transcendental realm (the Kantian thing-in-itself) when engaged in ingenious thought processes and when developing novel design patterns (see Mitcham, 1994: 29–33). Engineers are, in this Dessauerian sense, uniquely trained to maintain the relation between human cognition and the transcendental thing-in-itself in order to conjure more efficient objects into the world. A similar notion is found in physicist Freeman Dyson's widely cited statement that "Technology is a gift from God" and that "After the gift of life it is perhaps the greatest of God's gifts." It is not surprising that we should discover these quoted by the techno-enthusiast Aaron Bastani (2019: 31) who builds his claims on the notion that technological progress is "amounting to n othing more than an upgraded [more efficient] re-arrangement of previous information" (ibid. 63).

Instrumentalism has successfully infected popular perceptions on how the production costs of solar PV modules has dropped over the recent years. As I will explain in detail in Chapter 4, scientists and lay-people commonly evoke the "experience curve" to explain how and under what circumstances solar PV modules became cheap. This explanation, which is now at the cusp of common sense, is that solar PV cells have become cheap to manufacture thanks to the increased knowledge in how to manufacture them. Knowledge, in other words, increases production efficiencies, which in turn lower manufacturing costs. Thus, re-arrangement of previous information, to use Bastani's phrase, is sufficient for technological progress and for the realization of ambitious solar visions. Gone are concerns related to the necessary material prerequisites including wages and labor conditions, energy expenditures, land requirements, greenhouse gas emissions, and the extraction of the minerals.

Underscoring the instrumentalist preoccupation with efficiency is the very common but seriously underexamined assumption that engineering design and technological progress is "an effort (at first sight, of a mental sort) to save effort (of a physical sort)" (Mitcham, 1994: 221). The engineer, it is believed, solves "problems of fabrication that will save work (as materials or energy) in either the artifact to be produced, the process of production, or both" (ibid.). Design for efficiency, in other words, reduces physical expenditure through the application of knowledge. Or so the story goes. The notion that a cognitive activity could reduce physical expenditure is in fact no small assumption. A closer examination shows that it contradicts both the laws of thermodynamics and Epicurus's observation that "nothing comes from nothing" (for the former, see Georgescu-Roegen, 1975). It is worth mentioning that instrumentalist physicist Ernst Mach (1911: 49) explicitly criticized the principle of conservation of energy and argued that:

> What we represent to ourselves behind the appearances exists only in our understanding, and has for us only the value of a *memoria technica* or formula, whose form, because it is arbitrary and irrelevant, varies very easily with the standpoint of our culture.

That is to say, atoms, molecules, and physical events are semiotic representations, not true explanations, of the world (*epistemological idealism*). It follows that technology, as applied knowledge, must be considered as something immaterial.

From the standpoint of philosophical materialism, this implies that engineering invention and design can circumvent natural laws and cheat nature for the benefit of humans and other technology-wielding organisms. It follows that technologies do not require, so much as "save," matter-energy. This, I would like to argue, is the philosophical terminus of instrumental theory that is now at heart of the notion of "decoupling" that serves as a lodestar for ecomodernists. It is likely that this view also underscores the neoliberal commitment to take nature as purely instrumental and plastic for human benefit (cf. Pellizzoni, 2011). It follows, as Feenberg (1991: 6) argued, that there is little left but unreserved commitment to technology if we accept this essentially Promethean framework, which encourages the notion of technology as historical destiny.

Substantivism: A call for the sleeper to awake

If instrumental theory is the dominant theory of technology then "substantivism"[7] can be understood as its antagonistic rival. In contrast to instrumentalism, substantivist theories of technology point out that it is a dangerous fallacy to consider technologies merely as instruments without tangible consequences in the world. For substantivists, technologies are often catalyzers for wider social-political change. Since it is not widely recognized that technologies often encourage social change, substantivists view the uncritical acceptance of new technologies as something like a Trojan horse, which imposes often unwelcomed changes to social relations and values.

This perspective is primarily associated with Martin Heidegger and the French philosopher Jacques Ellul. In *The Question Concerning Technology*, Heidegger examines the instrumental notion of technology in order to understand its essence – "Enframing" (*Gestell*) – as a way of revealing nature as a "standing-reserve" of resources. The essence of technology, Heidegger contends, is nothing technological. Rather, we must understand technology as a historically specific method of revealing, or interpreting, what nature and history (or Being) is all about. Heidegger uses the case of the river Rhine and argues that the river is through the technological mode of revealing understood as a standing-reserve that subsequently appears to exist for the purpose of being exploited. The dam as technology, in other words, denotes not the dam itself with its turbines or valves, but the underscoring idea that the river exists primarily as an inert flow of exploitable matter (resource) available for human exploitation. In short, technology is more a phenomenological lens through which the natural world is understood and approached rather than itself a tangible material phenomenon in the world.

In a similar vein, Jacques Ellul (1964: xxv) defines technology as "the totality of method rationally arrived at and having absolute efficiency." The defining feature of the technological society as understood by Ellul is that every action and domain of social life are transformed by technology into rationalized processes with improved efficiency. However, technological values of rationality and efficiency occur at the expense of human values (connection, equality, sustainability, democracy, etc.). In the words of Langdon Winner (2020[1986]: 6), "technologies are not merely aids to human activity, but also powerful forces acting to reshape that activity and its meaning." Technologies thereby engender new lifeworlds.

Naomi Klein's view that solar power can evoke entirely new relations between humanity and the natural world comes surprisingly close to the substantivist view of both Heidegger and Ellul. As we have seen, Klein envisions that local communities can employ solar power as a means to claim independence from the dominant technological society, including its anti-social and anti-environmental tendencies.

But what are the ramifications for technological change according to substantivists? How would Heidegger and Ellul hold up against Klein's promising solar vision? For Heidegger, the essence of Enframing was inherited by technology from the modern physical theory of nature. An answer as to why Being is revealed through enframing can therefore only be found by questioning "the essential origin of modern science" (Heidegger, 1977[1954]: 23; cf. Merchant, 1980). According to Ellul, in contrast, technology has come to execute societal transformations autonomously. Whereas the technical operation is that of efficiency, the technical phenomenon is an autonomous consciousness in the presence of which humans are merely the "cellular tissue" in its total biology (Ellul, 1964: 142).[8] All societal contact with nature or history is mediated by technology that thereby functions as a barrier for authentic communication with nature. "Enclosed within this artificial creation" Ellul writes, "man finds that there is 'no exit'; that he cannot pierce the shell of technology to find again the ancient milieu to which he was adapted for hundreds of thousands of years" (ibid. 428). Thus, both Heidegger's and Ellul's views suggest that human-induced technological change needs to address deeper cultural issues than merely constructing solar modules from dismantled fossil infrastructure.

To break with the technological condition, if possible, humans in the technological society must awake from "technological somnambulism," a condition in which the symptom is to sleepwalk past the technological choices that produce the existence of the afflicted (Winner, 2020[1986]: 5–10). This implies that humans must become aware of what world they are making through their everyday technological choices and practices. Only once the end purpose of technology is made an explicit object of reflection can the appropriate means be discussed, developed, and implemented in possibly humane and democratic ways (cf. Illich, 1973). More importantly, since technologies may be inherently political, technological fixes are seldom the answer to social or ecological problems (Winner, 2020[1986]). Alternatives to technologies are therefore favored over alternative technologies (Winner, 1979; Heikkurinen, 2018). In relation to solar power, this suggests that communities should first articulate a broader human-environmental purpose derived from what Ellul would call "human values" and only then scrutinize whether and how the "technological values" of solar power fit into this purpose or not.

Critical theory: A political struggle over the technical code

The critical theory of technology is primarily associated with the philosopher of technology Andrew Feenberg (1991, 2008) who draws on perspectives from the Frankfurt School, Georg Lukács, and the early writings of Karl Marx. A key feature of the critical theory of technology is that it opposes the "take-it-or-leave-it attitude" toward technology that Feenberg believes characterizes the ying-yang dance of instrumentalism and substantivism. Critical theory moves beyond considering technology as something either emancipatory or repressive by drawing on constructivist technology studies that open for considerations of the role of social power in the design of technologies (Bijker et al., 1989; Feenberg, 1991).

More specifically, critical theory considers technological values (efficiency, productivity) as originating from the interests of the social group that has the most influence over the design process. Given the social character of the design process, technology is socially designed with a specific end in mind, rather than itself being an autonomous mind or cultural lens (as for substantivists). Feenberg (1991: 14) writes, "technology is not a thing in the ordinary sense of the term, but an 'ambivalent' process of development suspended between different possibilities." What technology is, in other words, depends upon what social class, or interest, is in control of the design. Technology is in this sense a plastic, malleable, phenomenon that is considered ontologically plural (i.e., it can be many different things). Following this reasoning, Feenberg considers technology a medium within which societal values and developmental pathways are politically negotiated and contested. What is contested, more specifically, is the "technical code," understood as "the realization of an interest or ideology in a technically coherent solution to a problem" (Feenberg, 2008: 52). Technology, in the most general sense, is therefore a mediator of social-political action and influence.

Critical theorists visualizes alternative technologies designed democratically, which can overcome the problems associated with capitalism, patriarchy, and modern industrialism. This requires active resistance to the current hegemony over the technical code through protests, grassroots movements, and reforms (see, e.g., Likavčan and Scholz-Wäckerle, 2018). However, before this can occur, the illusion of technological transcendence must be exposed as an instance of reification that operate to maintain social inequalities. Here, Feenberg (2008) draws on both Martin Heidegger and Herbert Marcuse and argues that the transcendence via technology is a cultural illusion at the heart of the modern experience that legitimizes divisions of labor.[9] Since no one is able to act without repercussions in a finite world, "technical action" represents not a full but *a partial escape from the human condition*" (Feenberg, 2008: 48, emphasis added). While technology provide humans with net benefits, it has adverse and unequally distributed impacts in the world for different social groups.

To understand how Feenberg's critical theory arrives at the plastic ontology of technology, we need to explore the underscoring "philosophy of praxis" situated within the Western Marxist philosophical tradition (Feenberg, 2014). In particular, since Feenberg draws heavily on Georg Lukács, we need to understand how Lukács approached the society-nature distinction and in what way this has colored the take on materialism in Feenberg's critical theory of technology. Lukács's seminal work *History and Class Consciousness* is perhaps one of the most influential texts for Western Marxists. In a famous footnote, Lukács limits the Marxist method to society and history while simultaneously levying a critique against Engels's dialectic method for claiming to know nature (Lukács, 1968[1923]: 24, footnote 6). To know anything about nature and matter, according to Feenberg's reading of Lukács, we would have to resort to investigating the *social production of nature* in which formulations of laws and ecological limits are cases of reifications in service of the capitalist class. Marxists such as John Bellamy Foster and Alfred Schmidt have rejected this approach as granting too much primacy to the realm of consciousness. They have consequently charged the early Lukácsian view with misinterpreting objectification (the coming to being, or evolution, of the natural world) as alienation, like Hegel (Feenberg, 2014: 124–128; Schmidt, 2014[1971]: 69–70; Foster, 2000: 244–249). Even Lukács himself, in what Feenberg calls a "unique example of philosophical self-misunderstanding," rejected his approach as a flawed attempt to "out-Hegel Hegel" (Lukács, 1968[1923]: xxiii; Feenberg, 2014: 126).

Despite this, Feenberg continues to develop Lukács's earlier statements on the society-nature distinction. According to Feenberg, Lukács's solution was to argue for two separate ontological realms in which the dialectic method was to be applied differently. That is to say, we cannot understand nature the way that we understand society (see also Burkett, 2013). This clarification is central because it shows that Feenberg's critical theory of technology is not seeking to understand technology from an interdisciplinary social-ecological perspective in which it is possible to study the common denominators of society and nature. While some (natural scientific) objects of thought exist independently of society, technology is not understood as such a natural object. To clarify, we can say that technology is more like the category "money" (semiotic) than the category "metal" (material) or "coin" (material-semiotic). This becomes clear if we remember that Feenberg's work concentrates on the process of design, thereby conceiving technology as primarily semiotic. It follows that technology is ontologically malleable and can be transformed through human praxis if only people were conscious that they themselves produced it (much like the category "money"). Technology, then, similarly to money, is a social medium through which relations are decided and orchestrated.

The notion that the natural (or physical) gives rise to the social (*ontological materialism*) is absent in this philosophy of technology. By following Lukács's earlier separation of society and nature only to exile technology to the social, Feenberg excludes the possibility that technology can be *both* a reification and an object in the world in the ontological-material sense (much like what is implied by the word "coin"). Notably, for Marx, even if human labor is taken away, "a material substratum is always left. This substratum is furnished by Nature without human intervention" (Marx, 1990[1867]: 133). "The physical bodies of commodities," Marx continues, "are combinations of two elements, the material provided by nature, and labor" (ibid. 133). Crucially, in Feenberg's critical theory of technology, the physical element of matter is missing, or at least appears to be underrepresented. This implies in turn that society and technology can be transformed from the inside (through design), without reference to an outside (or a human-environmental relation), much like a caterpillar metamorphoses into a butterfly in isolation.

Critical theory is a favored view among social scientists to understand the promises and pitfalls of solar power (see, e.g., Mulvaney, 2019; Newell, 2021). These social scientists acknowledge the unequally distributed harms and benefits of solar PV development and take seriously the environmentally harmful implications of green capitalism. After careful examination of these issues, however, they arrive at the conclusion that the harmful implications of solar power can be transformed through changes in ownership of the technology or through incremental improvements to the solar supply-chain. The harmful effects of solar power are thereby ethically softened with reference to a potential future condition in which the process of production has been overtaken by a working class organized for a different purpose other than capital accumulation. This technological optimism is evoked at the expense of a material view explaining how the present global environmental injustices emerge historically as a means to support industrial societies and the mass-production of commodities. I put this view to the test in this book as I seek to understand whether a world division of labor is not only present in, but inherent to, large-scale development of solar power.

Actor-network theory: Machines as social actors

The final theory reviewed in this chapter, Actor-Network Theory (ANT), has gained widespread popularity within the social sciences and sparked many controversies in recent decades. The main thinkers associated with ANT are Bruno Latour, Michel Callon, and John Law, two of whom have argued that ANT should not be understood as a theory at all (Latour, 1996: 377; Law, 2009). It is nevertheless true that ANT is made up of a set of principles and assumptions that together form a coherent "metatheory" (Bhaskar and Danermark, 2006; cf. Latour, 1996: 377–378).[10]

As a metatheory, then, ANT consists of a set of principles that are all constitutive of the central assertion of ANT; *that the world is exclusively made up of networks*. At heart, "ANT is a change of metaphors to describe essences: instead of surfaces one gets filaments [threads]," writes Latour (1996: 370). By substituting the metaphor of surfaces for the metaphor of network, ANT is seeking to get rid of the conceptual dichotomies (such as society-nature) that are believed to be at the root of the problems of the modern age (Latour, 1993, 2004). Nature and society are therefore shattered into millions of analytical pieces called "actants" that relate in complex "networks."

ANT invites us to think of the world in terms of actants and networks, but is technology an actant or a network? The question itself does not make sense within ANT because the actants are ontologically defined by reference to their relations in the network (Latour, 1988; MacGregor, 1991). This underscores the relation between humans and technological artifacts who are seen as mutually giving rise to one another: On the one hand, humans are in control of technologies as far as humans create and delegate tasks to technologies. On the other hand, technologies, such as a door-closer, "prescribe back what sort of people should pass through the door" and is therefore interpreted as a "highly moral, highly social actor" (Latour, 1988). As in substantivism and critical theory, technologies challenge human values. However, instead of thinking of technology as a consciousness or a politically contested medium for social transformation, Actor-Network theorists think of technologies themselves as social actors (Latour, 1988, 1996). As social actors, technologies are constituted through technology-human relations and are therefore not purely autonomous – but then again, neither are humans. This philosophy resembles postmodernist feminist Donna Haraway's categorization of the "cyborg," which blurs the boundary between the organic and the inorganic, much like the blurred boundary between the female and the male (1991).

ANT is frequently described as "material-semiotic" (Law, 1999, 2009; Law and Mol, 1995). This can be understood as a perspective that simultaneously takes into consideration the fact that materials (frequently referred to as "stuff") effect and gives shape to what is typically categorized as social. "The social," writes Law and Mol (1995: 276), "isn't purely social" but also material because all social relations involve relations with stuff. In turn, the same goes for technologies; "the electric vehicle is a set of relations between electrons, accumulators, fuels cells ... *and consumers*" (ibid. 276–277, emphasis added). In short, stuff is vital for the existence of people and people are vital for the existence of stuff. However, while networks involve and gives rise to material stuff, "a network is [itself] not a thing," writes Latour (1996: 378). Rather, networks are essentially semiotic. Networks, as opposed to actants, are invisible connections that are "immersed in nothing," writes Latour (1993: 128). In Chapter 4, we will see how ANT may have influenced one popular explanation of how solar power became cheap. This explanation focuses on how increased global "knowledge-exchange" has allowed for hitherto underdeveloped connections between social groups in the world and how they these new connections constitute a new network making solar power feasible.

From the position of philosophical materialism, some of the premises in ANT are based on fallacies. The first fallacy arises from the tendency to fetishize artifacts (Hornborg, 2017). Fetishization, in the Marxist sense, refers to the fallacy of assigning agency to commodities (Marx, 1990[1867]: 163–177). We will return to this point later. A second fallacy that ANT makes is to think of agency and relations (albeit not "stuff") as something essentially non-physical. Exponents of ANT (and new materialism) claim that inanimate things "have effects," "do things," "produce effects," or "have powers," and that these agential powers come from *within things*. For example, Bennett (2010: 18) writes, "so-called inanimate things have a life, that deep within is an inexplicable vitality or energy ... a kind of thing-power." This energy, or "thing-power," does not follow the regular habits of energy explained by physics.[11] The notion that things can animate themselves from within, without reference to an environment from which to draw the necessary matter-energy to exert their power, reduces relations in ANT to the study of signs (semiotics). Actants may be material "stuff" but relations are purely semiotic. This is so because the "effects" and "powers" that actants exert over one another are not understood as physical. Thus, Latour (1993: 378) writes, "what circulates [in networks] has to be defined like the circulating object in the semiotics of texts" and "a network is not a thing, but the recorded movement of a thing." Networks of signs are thereby thought to give rise to material "stuff" (*ontological idealism*). A third fallacy that indicates the ontological idealism underscoring ANT is the very notion that dichotomies can be transcended by efforts of thought alone. To simply un-think the dichotomies of the world is mistaken from a materialist point of view, since dualities are tangible, material, differences in the real world that arise historically. As shown by Adrian Wilding (2010), ANT thus cannot explain how the dichotomy that they criticize has come to bear significance in the world in the first place.

So what is technology? On lightbulbs as bright ideas and technology as a historical novelty

So, what have we learned from inquiring into these philosophies of technology? First, it is possible to see that there are both important differences and similarities across the four theories in the conception of technology. However, what is arguably most striking is the general commitment to understanding technology as ontologically non-material. The reference to human cognition, consciousness, design, and semiotic networks as core aspects of technology demonstrates the absence of ontological materialism. This is quite remarkable, but it should perhaps not come as a surprise in a culture that visualizes bright ideas as lightbulbs.

If these interpretations are correct, then we have every reason to question why ontological materialism is absent from contemporary philosophies of technology. The embryo of an answer can be found in Leo Marx's (2010) fascinating study on technology as a concept. While the concept of technology originates from a combination of the ancient Greek words *technē* and *logos*, it was not used in the now familiar sense of the word until well into the 20th century. Technology as a concept emerged as late as 1880 to fill the "semantic void" appearing due to considerable material changes in industrializing regions at that time. While the word technology first emerged to describe the complex material development of railway networks, the very "lack of specificity" made it "susceptible to reification" and eventually only the most obvious parts of the system stood in as "tacit referents" of the whole (Marx, 2010: 574). This suggests that the modern-day conception of technology has much more to do with the world-historical changes in the 19th century than with the ancient Greek notions of *technē* and *logos*.

Through a search on Google Book's Ngram Viewer we can confirm Leo Marx's assessment that the word technology is a historical novelty (Figure 2.1). This data shows how the word technology emerged in the beginning of the 1900s only to gain momentum by the 1960s and peaking in popularity just prior to the 2000s. Technique seems also to have entered the English language sometime in the beginning of the 1900s only to reach a modest peak in the late 1980s. In contrast, the word "machine" was used already from the 1700s onward and peaked just before the 1920s.[12] This confirms that the idea of technology is a very recent historical phenomenon. As we shall now see, reuniting the concept of technology with the specific socio-metabolic system emerging at this time in Europe and North America has major consequences for our understanding of what technology is.

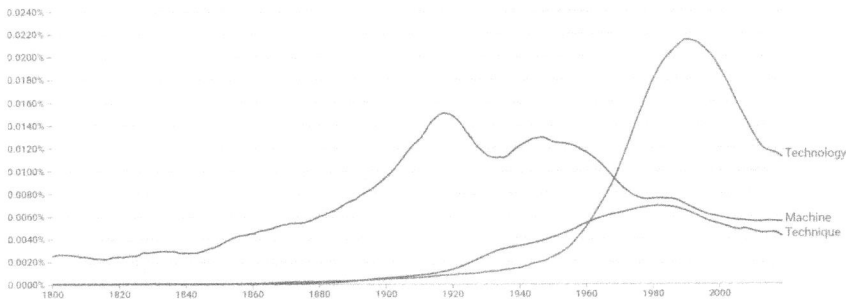

Figure 2.1 The occurrence of the words "technology," "machine," and "technique" in books published in the English language, 1800–2019.
Source: Ngram Viewer (2020a).

Foundations for a critical ecological philosophy of technology

I will now attempt to bring together Foster's devotion to Epicurus's and Marx's philosophical materialism and Hornborg's interdisciplinary understanding of technological systems. The aim is to point out how the above-mentioned philosophies of technology can be understood to represent different interpretations and analytical foci of the single social-ecological phenomenon of technology. The purpose is to offer a hypothesis of technology that is based upon the same philosophical assumptions of the world as research on the biosphere. The social-ecological philosophy that I propose is indebted to all the philosophies of technology reviewed above. However, based on its commitment to materialism, it is also distinct from them. It relates to the above-mentioned philosophies as follows:

- *Instrumentalism*: It agrees with instrumentalism that the truth-value of conceptions is determined through successful and sustained praxis, but adds that this truth-value should primarily be considered in terms of how such conceptions relate to social-ecological sustainability (i.e., the long-term survival of society in nature). For example, if the category of "technology" is part of a conceptual arsenal to exploit the natural world in an unsustainable fashion, then it can rightfully be considered to have a low truth-value and ought to be reconsidered.
- *Substantivism*: It agrees with substantivism that modern technologies may be indistinguishable from the social or political relations that they encourage or demand, but adds that such relations might also originate from the matter-energy required to produce them. For example, if solar PV projects require more matter-energy than can reasonably be supplied by a social group itself, then, in order for it to exist, it might encourage or demand specific social relations of exchange through which resources are appropriated.
- *Critical theory*: It agrees with critical theory that technological transcendence is an illusion of the capitalist mode of production but adds that technology is likely not transcendent under other modes of production either. For example, even if solar PV technology is installed to harness energy at low monetary costs, such harnessing is impossible without material demands on the natural world and cannot therefore transcend the fundamental metabolic condition of human societies.
- *Actor-network theory*: It agrees with ANT that modern technology can be defined by reference to a network but asserts that this network is foremost an international trade network orchestrating a global exchange of matter-energy. For example, even if solar PV technology design and manufacturing is orchestrated by actors in networks of knowledge-exchange, the production process of solar PV modules necessarily involves an exchange of matter-energy in international trade to take physical form.

To understand this philosophical position, and how it relates to solar power, we will first have to look to the notion of materialism and metabolism in Karl

Marx and understand how this position suggests an understanding of dialectics as metabolic.

Ontological materialism, metabolism, and dialectics

The nature of Marx's materialism has long been a matter of dispute (see Schmidt, 2014[1971]; Vogel, 1996; Foster, 2000). Despite wide-ranging disagreements, there is nevertheless a consensus that Marx's materialism was heavily influenced by ancient Greek and Roman philosophers such as Epicurus, Democritus, and Lucretius. Remarkably, modern understandings of space, time, evolution, and human origins were to a large degree anticipated by these philosophers. It is clear that Epicurus's philosophical materialism influenced not only Marx but played an extraordinary role for the founders of modern science and the English and French Enlightenment in general (Foster, 2000: 39–51). This hinged in large part on the fact that Epicurus's philosophy of nature was non-teleological. For Lucretius, the concept of nature was *mors immortalis* (immortal death), which refers to the inescapable and transitory mortality of nature itself. It was in the false notion that this condition could be escaped (e.g., the promise of an afterlife) that institutionalized religions could gain extraordinary powers. In opposition to this, Epicurus contended, true freedom was only ever achieved by embracing death as senseless (Foster, 2000: 6, 36).

In this material philosophy, dynamic and open-ended changes in nature take precedence over God or final causes. By extension, humans are not created in the image of God or Spirit but are temporal and sensuous beings through whom nature actively engages with itself. Thus, Marx (2000[1841]: in Ch. 4) quoted Epicurus: "In hearing nature hears itself, in smelling it smells itself, in seeing it sees itself." Contrary to Lukács's charge against Engels that humans cannot know nature dialectically, Epicurus's and Marx's position implies that nature can be known, because knowing nature is synonymous with knowing oneself. This does not mean that human knowing is always true or complete. Rather, as a biological species, *what is true is whatever mental representation of the world sustaining a given human organization (society) over time in any given environment (nature)*.[13] In this, signs are as much supportive of material relations as they are "plays," "struggles," "quests for mastery," or whatever signs or meaning the social metabolism invites or demands (cf. Rappaport, 1968; Harris, 1980).

If critics argue that such epistemological materialism is Baconian, it is because they fail to acknowledge the relationism in Marx's notion of metabolism (Foster, 2000: 10–11). Schmidt (2014: 78) illuminates this contested topic by noting that Marx abandoned his early Baconian view once he replaced the linear notion of human appropriation of nature with the dialectical notion of "metabolism" [*Stoffwechsel*]. Stoffwechsel, or metabolism from the Greek *metabolē* (exchange), was a term that came to be used by German biologists in the 1800s to explain how cells in the human body could maintain their material form over time. The understanding that there was a similar metabolic exchange between human bodies and their environment was later pointed out by Justus von Liebig. Influenced by Liebig, Marx wrote, "man lives on nature ... nature is his body, with which he must remain in continuous exchange if he is not to die" (Marx, 2000[1844]: 31). Crucially, since human organisms are social, the metabolic exchange also applies to different social formations. Hence, some interdisciplinary scholars use the term "social metabolism" to denote the socially organized relation to the environment (see 'metabolism' in Glossary). A cell, a human or a human society, is then, primarily, a materially integrated component of nature sustaining itself through metabolic relations with its surrounding. The essential difference between the Baconian view and the metabolic view is whether nature (the biogeochemical processes in the biosphere) is understood as an external resource existing for human exploitation or as the necessary counterpart of society without which society would not exist.

The notion that human societies can escape this metabolic relation to nature, bound within *mors immortalis*, is a central theme in religion and political ideologies throughout history. Today, in both instrumentalism and ANT, we see attempts at breaking with this fundamental metabolic condition in two opposite ways. While instrumentalism and ecomodernists champion a radical separation of society and nature, ANT champions a radical unity of society and nature. Neither approach is correct from the metabolic view because metabolic exchange forms a relation that is characterized by both connection and separation. While society and nature are the same at one level, they cannot exist in unity without a separation that facilitates a metabolic exchange between the two. We might say that relations necessitate separation; otherwise, there would be nothing to connect. This is true if we consider human-to-human relations in our everyday life and it is true physically, as becomes evident by the "useful fact" that the "universe is not one solid mass, all tightly packed," as Lucretius (2007: 18) wittingly observed. This means that we have good reasons to question whether escaping from the social metabolic condition through solar power or any other technology or method is possible. The material perspective of Epicurus and Marx implies that this is indeed an impossibility. This is something that we learn also from thermodynamics.

Thermodynamics, evolution, and exosomatic organs

The modern science of thermodynamics has its antecedents in 19th-century Britain and France. At that time, the rise of the political power of the bourgeoisie was increasingly connected with the steam engine, employed to pump water out of coalmines and perform mechanical work in industries (Malm, 2016). Still, the early steam engines managed to convert only a paltry 2% of the potential energy in coal into useful work. Increasing the efficiency of steam engines was therefore a key concern for early industrialists. At this time, one important question was how efficient steam engines could become. Was it possible that steam engines could be developed to feed on their own boilers in perpetuity without further human effort? These questions were intimately bound to relations of power. As Rabinbach (1992: 58) argued, the quest for perpetual motion was "the phantasmagoria of a society dedicated to making work superfluous." In effect, there was a "pervasive moral criticism of those who resisted work" because they were defectors in the "search for an alchemy of work without struggle" (ibid. 58). The commitment to an unrealized techno-utopia void of metabolic implications (e.g., work without struggle or technology without environmental impact) is reminiscent of both instrumentalism and critical theory.

It was in the search for a *perpetual motion machine* that Sadi Carnot discovered the irreversibility of heat passing from hot to cold, now known as the second law of thermodynamics (the entropy law). The implications of Carnot's engine were later formalized by Rudolf Clausius (1867), who stated that the transfer of energy from a warmer to a colder body always implies a total loss of useful energy. The nature of energy, which Hermann von Helmholtz (2001[1862]) considered a universal indestructible "Kraft," was that it universally tended toward less useful states. It slowly dawned; the world was characterized by entropy, an inescapable tendency toward disintegration and thermal equilibrium. The cornucopian potential in the first law of thermodynamics – that energy cannot be created or destroyed – was effectively shattered with the understanding of the entropy law. It implied that the omnipresent "Kraft" could not be reused infinitely. The implications were game-changing because they struck at the core of the 19th-century European cosmology.

As Stokes (1994: 67) points out, "the effect of the thermodynamic laws on the thinking about evolution in the universe was profound." How could organisms live, grow, and evolve in a universe characterized by entropy? It did not take long until the biologist Herbert Spencer (1904) provided an answer: The human body counter-maneuvered the law of entropy by drawing energy from its environment. Schrödinger (1945: 75) later defined life in general as that which "feeds upon negative entropy":

> Thus, the device by which an organism maintains itself stationary at a fairly high level of orderliness … really consists in continually sucking orderliness from its environment. This conclusion is less paradoxical than it appears at first sight. Rather could it be blamed for triviality.

However, it is through this "triviality" that the anti-mechanism of the law of entropy becomes evident, since it implies that nothing can be said to exist simply by reference to its internal parts. For example, take a human body, a tree, or a clockwork mechanism. A human cannot continue without food; a tree cannot continue without sunshine; a clockwork mechanism cannot continue without a human winding up its spring. The entropy law was and still is proof of the inescapably relational character of artifacts, life, and societies (Georgescu-Roegen, 1971; Bateson, 2000[1972]).

In terms of evolution, humans have been highly successful in maintaining metabolizing collectives in a wide range of environments without any drastic variation in their physiology (Bates, 2001). This can partly be explained by the fact that human organs are not all part of the physiology of the individual human body, as becomes evident through thermodynamics. Apart from the "endosomatic organs" that are part of the body, human depend extensively on "exosomatic organs" outside their bodies that provide access to a range of different environments (Lotka, 1956). One universal example is fire, an exosomatic organ for digestion and an aid to making environments more accessible to the human body through cooking and more effective hunting. Another example is the plow, an exosomatic organ intensifying the amount of humanly available biomass that can be extracted from a given environment. Yet another example is the British Imperial coal network and the colonial triangular trade that facilitated an appropriation of labor time and natural resources from ever more remote environments and peoples (Pomeranz, 2000). As will be elaborated below, all such organs provide matter-energy, *but only through prior and continual use of matter-energy* – a point unanimously overlooked or downplayed in the philosophies reviewed above.

Even if exosomatic organs all in varying degrees facilitate an appropriation of resources from the environment, different organs may imply different relations within a society. Lewis Mumford (1964) sheds light on this issue by separating what he calls "democratic technics" from "authoritarian technics." Examples of democratic technics are fire, baskets, nets, bows, and simple water pumps, all defined as democratic with reference to the fact that they can be learned, produced, and controlled by any adult member of the species. Democratic technics, he writes, are "relatively weak, but resourceful and durable" and work best in contexts where aspirations for accumulation are low (Mumford, 1964: 2). However, as is evident both historically and in our own time, social aspirations may imply material pressures on environments that exceed their regenerative biocapacity (Pomeranz, 2000; WWF, 2020). For such aspirations to be saturated, energy and material resources have to be extracted from non-local environments and peoples (Hornborg et al., 2007; Rice, 2007). The "authoritarian technics" required for such aspirations included an orchestration of both nature and people in systems of material and ideological power (Mumford, 1954). In short, these organs cannot be democratically produced or maintained since they necessitate (and in a sense are) undemocratic relations of production whereby some people work for the benefit of others.[14] Mumford (1964) pointed out that contrary to the humble democratic technics, authoritarian technics excel simultaneously in both mass construction and mass destruction. Whether these technics, or organs, are understood as emancipatory or destructive is therefore contingent upon how particular social groups are positively or negatively affected by them, something that today often corresponds to particular geographical locations (see Hornborg, 2014).

Ecologically unequal exchange and machine fetishism

Like Mumford's "authoritarian technics," Alf Hornborg's interdisciplinary work on ecologically unequal exchange has suggested that modern technology is inseparable from the global social-ecological arrangement orchestrated by European colonial powers at least since the 18th century (Hornborg, 1992, 2001, 2016).

Drawing on world-systems analysis, dependency theory, and ecological economics, the theory of ecologically unequal exchange proposes that capitalist world trade orchestrates an exchange whereby richer (core) regions of the world appropriate resources from impoverished (peripheral) regions of the world (Giljum and Eisenmenger, 2004; Jorgenson et al., 2009; Dorninger and Hornborg, 2015; Dorninger et al., 2021). It explains these differences by empirically demonstrating how economic exchange – conventionally measured in money – facilitates an unequal exchange in terms of labor time, embodied land, and/or natural resources. This is possible because any *symbolically equal* exchange, for example in terms of money ($100 for $100), may simultaneously imply a *physically uneven* exchange in terms of resources or resource investments (say, 100 kg for 10 kg). Hornborg (1998: 131–132) argues, "*market prices* are the specific mechanism by which world system centres extract exergy from, and export entropy to, their peripheries."[15]

According to Wallerstein's world-system analysis (2004, 2011a–d), the modern world system, more commonly known as capitalism, emerged historically as a world division of labor which can be analytically separated into three regions; the periphery, the semi-periphery and the core.[16] Core regions in the world economy are those in which the production process is controlled by corporations backed by strong states. Through tariffs, subsidies, tax benefits, and other measures, states in core regions facilitate a wide margin between production costs and sales prices for corporations. Thus, wages and prices remain high in core regions because market competition is effectively annulled. In contrast, corporations in peripheral regions of the world economy do not enjoy the same state benefits. They are therefore "truly competitive" in a manner that pushes down wages and prices on their products (Wallerstein, 2004: 28). Semi-peripheries exhibit both core-like and peripheral-like conditions of production.

If we look at China as an example, wages and prices remain low in China by international standards even if there is a growing middle class enjoying a higher income. When the country opened for international capital in 1978, the inflow of foreign direct investments (FDIs) forced state-operated firms to compete with private firms from overseas. This led the Chinese state to purge the state-operated firms from low productivity workers in the 1990s (Yang et al., 2010). These included less educated workers, old workers, and women in precarious positions. As production was made more efficient, wage levels increased. State regulations thereby facilitated higher wages. Even as wages increased, however, the unskilled population was left without a steady income, effectively becoming, in Marxist terms, a "reserve army of labor," desperately offering unskilled labor at very low wages (Yang et al., 2010). This process continues today. While wages and worker benefits are increasing in China, these effects are unevenly distributed within the nation. When commodities are exchanged on the international market under the resulting price differences, a net transfer of embodied resources tends to flow from periphery to core, periphery to semi-periphery, and semi-periphery to core. By extension, much of the environmental load of the high-consumption lifestyles in the core is displaced to the world peripheries (Hornborg, 2009; Dorninger et al., 2021). In Chapter 4, I focus on understanding how Germany's solar development was contingent on environmental load displacements to China in the rise of solar power.

According to the theory of ecologically unequal exchange, technologies have been intertwined with socially determined rates of exchange (prices) whereby some nations have been able to appropriate resources from other nations to build and maintain modern infrastructure at least since the mid-19th century. Hornborg (2006) shows, for instance, that in exchanging British cotton manufactures for North American raw cotton at equal monetary prices, Britain in 1850 established a net flow of embodied labor and embodied land to Britain. Technology can be understood from two vantage points in the orchestration of such material exchange:

1   Machine technology such as water mills and steam engines *necessitated* a concentration of resources to be constructed and operated. For example, the development of early industrial machinery in Britain was based upon the import of large amounts of Scandinavian iron, and later, after the repeal of the Corn Laws, Russian and Prussian wheat for feeding the labor force (Pomeranz, 2000; Debeir et al., 1991: 108–111).

2   Machine technology such as water mills and steam engines *facilitated* asymmetric metabolic relations by lowering production costs in the cotton industries. This in turn leads to more favorable exchange rates and further rounds of resource appropriation. In such a way, not only state power, but also productive infrastructure contribute to low production costs relative to commodity prices (Warlenius, 2016). To this rationale, we add the use of military technologies to subdue and exploit peoples around the world to secure labor and resource-abundant or geopolitically favorable locations (Headrick, 2010).

Asymmetric metabolic relations, in theory, are what modern technology at once necessitates and facilitates. As such, modern technology can be understood as emerging due to the ecologically unequal exchange that allowed the elites of the British Empire to eventually accumulate more resources than the biocapacity of the British Isles could provide. The same asymmetric relations, Hornborg contends, can be observed in the distribution of light in night-time satellite images of the Earth today. If the "inventors of nuclear bombs, space rockets, and computers are the pyramid builders of our own age" (Mumford, 1964: 5), then Hornborg argues that ecologically unequal exchange is the indispensable mechanism through which the necessary labor and natural resources for these inventions can be accessed.

The notion that modern technologies necessitate ecologically unequal exchange has profound implications for a philosophical consideration of technology. This is so because the work that technological artifacts appear to perform in a local environment necessitate resource expenditures (materials, land, labor, etc.) elsewhere in the world economy. Theoretically, this is equally true for smartphones as for solar PV modules. This means that "the rationale of machine technology" is not necessarily to do work, but rather to "(locally) save or liberate time and space, but (crucially) at the expense of time and space consumed elsewhere in the social system" (Hornborg, 2006: 80). Solar PV modules provide obvious benefits (time, energy "saved" or "harnessed") for those who can afford them, but they simultaneously imply obvious burdens (time, energy "spent") across the global production process. From this, we may ask, do the physical costs shouldered by nature and workers outweigh the physical benefits gained by the technology user? With reference to the second law of thermodynamics, the transformation of matter-energy is always accompanied by an increase in disorder. This alone excludes the possibility that technology is something that delivers net benefits in a physical sense. In addition, as pointed out by Lucretius (2007: 11), "nothing can be made from nothing." Technologies, then, do not add anything physical to the world, merely rearrange and transform already existing matter-energy.

From a critical ecological perspective, to believe that technologies provide physical net benefits in the world is an illusion maintained by the fact that the adverse costs of any given technological artifact or efficiency improvement are displaced to nature or to other parts of the world. The question, then, is for *whom* a given technology is biophysically worthwhile. In the world economy, the burdens, or costs, associated with technology are taking place far from the everyday sensuous experience of the user. This is certainly true for solar PV modules, as we became familiar with in Chapter 1. The separation of production and installation is arguably the root of "machine fetishism" wherein technologies appear to have innate productive qualities (or, agency) since they are thought of as isolated from the global social-ecological arrangement that generated them (Hornborg, 1992, 2016). Rather than having innate productive qualities, however, technologies are better understood as having productive qualities due to resource expenditures elsewhere in society or nature. Solar technology, in this view, is thought of as productive because it has implied resource expenditures elsewhere.

The "agency" or "thing-power" spoken of by new materialist and ANT theorists is therefore not innate to the technological artifacts themselves but granted to them by virtue of being the embodiments of resources dissipated elsewhere. To put it simply, we can say that the solar module is working because it has implied a loss of resources (low entropy) earlier in its life cycle. The expenditure of low-entropy resources, which is often associated with expended labor time and environmental impacts, is therefore inherent to the function of solar technology. It is not, as instrumentalism and critical theory would have it, an evil external to its operation that is possible to banish with incantations of efficiency improvements or ownership transformation. The degree to which a given technology works to the maximal benefit of the user depends upon to what degree the loss of low entropy can be displaced to other systems (social or natural) or not. The question concerning technology is therefore ultimately a matter of matter-energy distribution (not addition) between natural processes (e.g., nutrient cycles) and social groups.

In sum, the critical ecological framework in this book is underscored by at least six basic assumptions. These assumptions are all related to the overarching assumption explored in this chapter, i.e., that philosophical materialism provides invaluable insights into the processes of nature and history. This includes the following:

- First is an agreement with ontological materialism asserting that human populations emerge from nature's independent processes, to which they therefore are metabolically bound. From this assumption emerges an understanding of the fundamental paradox that all human-environmental relations imply human-environmental separation.
- Second is an agreement with epistemological materialism when assuming that human semiotic representations of nature are true to the extent that they support a particular metabolic interaction with – and so survival in – nature. Given today's ecological problems, this assumption motivates a questioning of the modern outlook of science, economics, and technology and their relation to the metabolic reliance upon fossil fuels and the Industrial Revolution (for science and economics, see Georgescu-Roegen, 1971; Daggett, 2019). The notion that technology constitutes a problematic concept in modern culture is reflected in the lack of ontological materialism in contemporary philosophies of technology that have effectively omitted biophysical nature.
- Third, any given technology is part of a socially organized metabolism.
- Fourth, since "nothing comes from nothing" and "everything must go somewhere," technologies do not add – merely redistribute and dissipate – matter-energy in the world. By virtue of being made of large amounts of initially dispersed material compounds, modern technologies require global social relations that concentrate resources. The neglect of this fact leads to the pervasive modern cultural misrepresentation of technologies referred to as "machine fetishism."
- Fifth, the question concerning technology is foremost a question of matter-energy distribution across social groups and natural processes in contrast to the conception of it. This is the theme of this book, as applied to solar power.
- Sixth, in line with practical materialism (see the definition above), human-environmental relations can change in a more sustainable direction through deliberate human technological practices adjusted to carrying capacities and nature's processes (see below).

The technological continuum

To explain my critical ecological approach to technology, I propose a continuum of technology (Figure 2.2) that analytically divides the complete phenomenon of technology into:

1 Technology as **past** social conditions and consequences (prior to being assembled as artifacts).
2 Technology as an artifact in the **present**.
3 Technology as **future** social relations and consequences.

PAST Socio-ecological prerequisites (impacts, relations) PRESENT Socio-ecological consequences (impacts, relations) FUTURE

TIME

Here technologies appear as independent objects

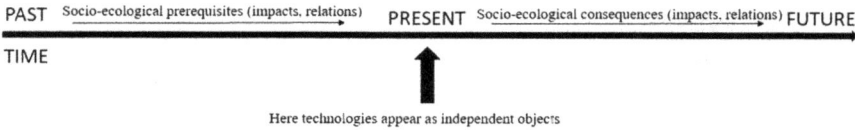

Figure 2.2 The technological continuum, representing an exchange of embodied matter-energy from the left end to the right end. Reprinted with permission from Oxford University Press. Original image can be found in the book *Sustainability Beyond Technology: Philosophy, Critique, and Implications for Human Organization*, edited by Pasi Heikkurinen and Toni Ruuska (Roos, 2021).

With reference to this continuum, Hornborg's concept of "machine fetishism" explains a collective difficulty in thinking of past social-ecological relations and consequences – and so the full material continuum – as essential to what technology *is*. If the full continuum is taken into consideration, we understand that we are dealing with a relation between the systems on the left side of the continuum (among which the loss of low entropy occurs and the technological artifact's capacity to do work is generated) and the systems on the right side of the continuum (in which technological artifacts are applied for their capacity to do work). In this view, technological artifacts in the present embody low entropy (or resources) "spent" that can be put to work by the user to facilitate further rounds of appropriation in the future. In other words, technologies are means for continuous ecologically unequal exchange (Hornborg, 2003).

Let us briefly take a closer look at the biophysical past of a specific technological artifact to see what is omitted in the fetishized perspective. To this end, I would like to refer to the German company Nager IT, which produces the "Faire Maus." The aim of the company is to produce a computer mouse that is a hundred percent fair. It meticulously documents the working conditions under which all the mouse's material components are manufactured. Figure 2.3 is a summary illustration of the company's effort to trace and assess the mouse's commodity chain. This represents the biophysical past of a particular technological artifact – in this case a computer mouse – as an illustration of the technological continuum. The mouse itself, of course, represents the technological artifact, such as it appears to consumers in the present. In this illustration, we can see how the biophysical past of a modern technological artifact is immensely complex. To produce something as seemingly simple as a computer mouse requires more than 20 material components extracted from around the world, which are then transported, processed, packaged, and assembled in a range of different stages. If the boxes in the illustration represent particular material components or manufacturing processes, the lines in between the boxes represent instances of matter exchange and transportation.

To produce a solar PV module, including the necessary inverter, performance monitor, and battery or grid connection, involves even more complex commodity chains. Very few have attempted to account for the energy required in these

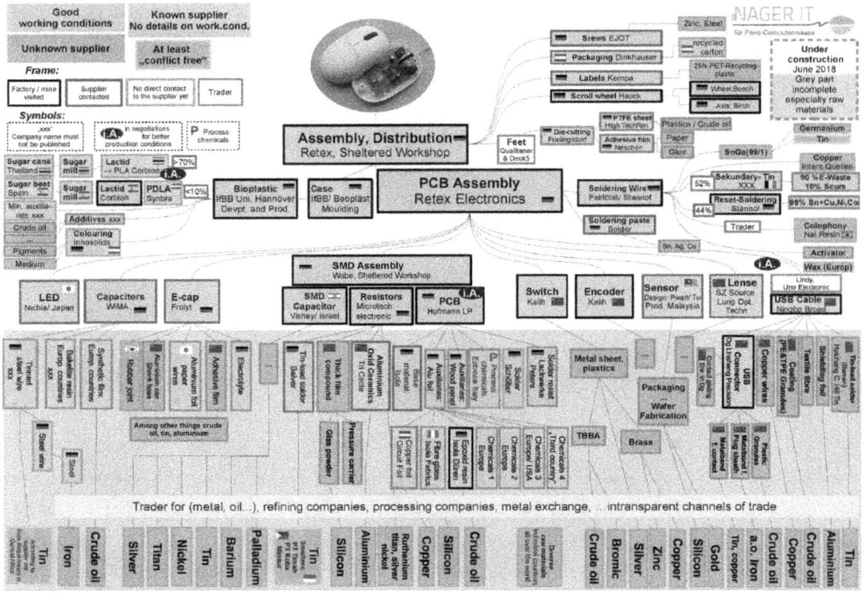

Figure 2.3 The commodity chain of the computer mouse "Faire Maus." Produced by the German company Nager IT. With permission from NAGER IT (https://www. nager-it.de/en).

commodity chains (for one example, see Prieto and Hall, 2013). A recurrent problem for ecological economists who are interested in the exchange of matter-energy is the fact that commodity chains are conventionally accounted for in terms of money. Money, however, is simply the symbolic representation of what is actually being exchanged (matter-energy) and spent (low entropy). In this physical sense, technological artifacts do not come from money, but from matter-energy that has been appropriated from nature and transformed by labor. In industrial capitalism, this process occurs with the aid of highly energy-dense fossil fuels (see Chapter 3). So far, it seems, this fundamental biophysical condition of technological artifacts has been sidestepped in the philosophies of technology reviewed above.

The review of the assumptions in contemporary philosophies of technology suggests that it is common to consider technology as something ontologically non-material, yet often with tangible consequences in nature. This is most obvious in instrumentalism, in which technology, like a *deus ex machina*, is lowered down onto the theater of the present to solve material problems for the future. This is the view that powerful organizations such as the Organization for Economics Co-operation and Development (OECD) and the United Nations Environment Programme (UNEP) are now operating with, using defunct notions of "green and sustainable growth" (OECD, 2011; UNEP, 2014; cf. Hickel and Kallis, 2019). This omission can also be deduced from the assumptions in ANT, which thinks of technologies as emerging from semiotic networks (as if that which the boxes and lines of the Faire Maus commodity chain illustration represent are simply ideas). The reference to technology as design or consciousness is similarly problematic since it downplays and sometimes obfuscates the social-ecological origin of technological artifacts. This is not to say that design and consciousness are not necessary aspects of technology. However, it does mean that design and consciousness are not sufficient to describe the complete material phenomenon of technology. To acknowledge the complete technological continuum implies a consideration of what a given technological artifact at once *necessitates* (the left side of the continuum) and *facilitates* (the right side of the continuum). By understanding technologies this way, it becomes possible to evaluate how a given technology alters nature's processes and/or contributes to a transformation of human-environmental relations within the confines of the biosphere. The rest of this book will focus on understanding more precisely how large-scale development of solar PV technology *necessitates* and *facilitates* environmental load displacements through the world division of labor articulated in world-system analysis.

Perhaps the most important point offered in this chapter is that technologies should not be assumed to alleviate environmental pressures in one area without implying such pressures elsewhere in the world.[17] With that said, the notion that modern technology is a means to orchestrate unequal exchange of matter-energy does not mean that technology can or should be rejected. As we have seen, all organisms, or collective of organisms, require metabolic strategies for their very survival, i.e., strategies to suck orderliness from their environment. Without such strategies, they would cease to exist. This is an inescapable condition of being an organism, or a collective of socially organized organisms, in nature. While this appropriation is by definition unequal, human organizations can, in theory, choose to what degree they expend other people's resources by exploiting their land and labor. As current global ecological footprints greatly exceed the carrying capacity of the Earth (WWF, 2020), this will be a major challenge for 21st-century societies seeking to transition away from fossil fuels in a just and humane way.

Gender, race and ecology in technology

To substantiate the critical ecological philosophy of technology, it is necessary to include thoughts on how gender and race are implicated in modern technology. In this book, I have chosen to focus on the contrast between vision and reality. What such visions and realities entail should arguably be broadened to not only include considerations of environmental sustainability.

Regarding issues of gender, the perhaps main question is whether patriarchy – understood as a male-female relation of male domination with consequences for the unequal distribution of benefits and burdens – is a necessary condition for modern technology based on ecologically unequal exchange. In relation to solar power, research suggests that the negative effects of large-scale solar PV parks disproportionally affect women (Stock and Birkenholz, 2019). This is an important finding. But to ask whether specific gender relations are necessary for solar PV development means looking beyond the social-environmental impacts associated only with the latter stages of the commodity chain (including installation and usage). It includes an eco-feminist lens examining the lived conditions and social relations of patriarchy as playing a part in solar PV development prior to the assembly of the modules. According to Carolyn Merchant (2014: 63), such a lens is found among socialist feminists, who

> see nonhuman nature as the material basis for human life, supplying the necessities of food, clothing, shelter, and energy. Materialism, not spiritualism, is the driving force of social change ... Socialist feminism views change as dynamic, interactive, and dialectical, rather than as mechanistic linear, and incremental. Nonhuman nature is dynamic and alive.

The philosophical materialism underscoring this position is necessary for understanding the relation between gender and ecologically unequal exchange. Remembering the previous distinctions between the categories metal (material), money (semiotic) and coin (material-semiotic), we may ask to what degree the social construction of "gender" is similar to the social construction of "money" in making modern technology possible by facilitating uneven distributions of resources and burdens. The two are of course interlinked. It is no secret that women are frequently paid less than men for the same work. This simple fact reveals the potentially pivotal role that patriarchy plays for the uneven valuation of labor and resources that determine ecologically unequal exchange. One central task for furthering the theory of ecologically unequal exchange and its role for solar PV development is to link it to what Mary Beth Mills (2003) calls a "gendered global labor force." The analytical focus synthesizing ecology and gender would then focus on how the globally uneven valuation of work and resources – which is here theoretically necessary for modern technology to exist – may be reproduced through immaterial gender categories.

Focusing on race may also help illuminate the cultural categories and social relations necessary for modern technological infrastructure. As I will demonstrate in Chapter 3, the historical emergence of modern technology is largely indistinguishable from the European colonialization of non-white peoples throughout the world. The semiotic representation of different social groups in the world economy has played a decidedly central role in exploitation of non-European peoples. In Chapter 5, I show how the conventional definition of technology is so narrow that it excludes how solar PV development implicates the exploitation of non-white and non-European peoples in the world economy. If such exploitation is not included in the definition of technology, then solar technology appear as a more reasonable solutions to issues such as social injustice and ecological sustainability. Such instrumentalism, as we have seen, encourage the problematic notion of a pre-determined, God-given, technological progress in human history. This, in turn is linked to the fallacious notion of "racial progress" as explained by Seamster and Ray (2018). According to these authors, industrial capitalism operates with the Enlightenment belief that racial change is inevitability progressive. Along with drawing narrow technological boundaries, this teleological notion has a tendency to delegitimize research and practices for more radical social transformations necessary for racial justice. This commonality provides fertile grounds for further considering how teleological notions of race and technology are implicated in solar PV technology.

In sum, the critical ecological philosophy of technology offers a philosophical foundation rooted in materialism which makes it possible to distinguish between semiotic representations and physical reality. Its designated purpose is to understand how unsustainable relations of power and environmental harms are legitimized – and so reproduced – through false categories and erroneous social visions. This purpose is not confined to considering the ecology of technology, simply because the representations and realities of gender and race are likely central for ecologically unequal exchange and the existence of modern technology.

Notes

1 This is a reworked version of a chapter published by Oxford University Press in the book *Sustainability beyond technology: Philosophy, critique, and implications for human organization*, edited by Pasi Heikkurinen and Toni Ruuska (Roos, 2021).
2 Kant's notion was preceded by the ancient Greek philosopher Epicurus's understanding of nature as *mors immortalis* (see below). Notably, the Greek word *Physis* means nature, such as it "creates itself" and "emerges from out of itself" (Feenberg, 2009: 160). The Chinese word Ziran similarly means, "it is so by virtue of its own" (Chan, 2005: 542, emphasis in original).
3 Notably, Marx's and Engels's program, historical materialism, was an attempt to understand the evolution of human organization in this sense.
4 Here I have purposefully left out the question of whether consciousness (mind) is dependent or independent from nature. I address this issue in more detail below.
5 I distinguish between technological artifacts (or objects) and technology as a continuum wherein technological artifacts are temporal manifestations (see below under sub-headline "The technological continuum: An illustrative model"). In this view, a machine is a type of technological artifact.
6 Vernadsky's (1945) notion of the human brain as the determinant of technological progress is an example of this view.
7 Substantivism is sometimes called "the cultural approach" (Drengson, 1995: 39–50).
8 In more recent work, this consciousness is not so much autonomous as something of "our own subconscious intelligence" (Drengson, 1995, quoted in Heikkurinen, 2018: 1659).
9 Feenberg's critical theory shares some philosophical assumptions with both instrumentalism and substantivism. Feenberg agrees with instrumentalism that the ontology of technology is decided in the process of its design but rejects the notion of its mysterious origin or purpose. Feenberg also agrees with substantivism that technological values must be opposed and questioned but rejects the notion of technology as in any way inherently political or problematic.
10 These include (i) a radical rejection of conceptual dichotomies, (ii) the principle of symmetry, (iii) a definition of agency as "having an effect," and (iv) the principle of decentralized power.
11 MacDuffie, in his study on the fictionalization of energy, shows that energy as a metaphor in this sense arose from the erroneous 19th-century British experience of the city as a closed system capable of feeding on itself (MacDuffie, 2014).
12 In other European languages, the results are similar. In French, the word "technique" shows a similar curve as the word "technology" in the English language (Ngram

Viewer, 2020b). In German, the word "technologie" follows a curve very similar to English (Ngram Viewer, 2020c).

13 For example, as Foster (2000: 55) noted, "in Epicurus is found even the view that our consciousness of the world (for example, our language) develops in relation to the evolution of the material conditions governing subsistence." God, in this sense, can be true (but not real) if the specific practical actions derived from the worshipper positively affects the reproduction of the social metabolism. In contrast to the instrumental conception of objective truth as socially constructed, human knowledge of the world is here understood to operate in feedback with a real material world (Rappaport, 1968; Bateson, 2000[1972]).

14 According to Engels (1972[1874]), industrialism and machinery is inherently authoritarian irrespective of who owns the means of production. In effect, "wanting to abolish authority in large-scale industry is tantamount to wanting to abolish industry itself" (ibid. 731). Notably, Engels did not argue for the abolition or moral questioning of industrialism and machinery because he saw these developments as historically inevitable.

15 For a description of "exergy," see the Glossary.

16 Such a definition of capitalism differs from positions that considers institutionalized wage labor, class-based ownership of the means of production, or even the combustion of fossil fuels as core defining features of capitalism. The famous Wallerstein-Brenner debate signifies a core dispute regarding the scale of the level and unit of analysis for understanding capitalism, where Brenner has challenged Andre Gunder Frank's and Wallerstein's analytical focus on exchange and class relations at the level of the emerging colonial world economy (Brenner, 1977). I have sided with Wallerstein, who provides a basis for considering the rise of industrialism as fundamentally intertwined with world strategies for environmental load displacement prior to the fossilization and mechanization of the industrial manufacturing process (for this position, see also Denemark and Thomas, 1988; Bunker and Ciccantell, 2005; Barca, 2011).

17 Similar forms of displacement may also apply to the working conditions and gender relations altered when new technologies are commercialized in the world core.

References

Barca, Stefania. 2011. Energy, property, and the industrial revolution narrative. *Ecological Economics* 70: 1309–1315. https://doi.org/10.1016/j.ecolecon.2010.03.012.

Bastani, Aaron. 2019. *Fully automated luxury communism.* London, England and New York, NY: Verso.

Bates, Daniel. 2001. *Human adaptive strategies: Ecology, culture, and politics.* Boston: Allyn and Bacon.

Bateson, Gregory. 2000[1972]. *Steps to an ecology of mind.* Chicago, IL and London, England: The University of Chicago Press.

Bennet, Jane. 2010. *Vibrant matter: A political ecology of things.* Durham and London, England: Duke University Press.

Bhaskar, Roy. 1991. Materialism. In *A dictionary of Marxist thought*, edited by Tom Bottomore, Laurence Harris, V. G. Kiernan, and Ralph Miliband, 369–373. 2nd ed. Oxford, England: Blackwell.

Bhaskar, Roy, and Berth Denermark. 2006. Metatheory, interdisciplinary and disability research: A critical realist perspective. *Scandinavian Journal of Disability Research* 8(4): 278–297. https://doi.org/10.1080/15017410600914329.

Bijker, Wiebe, Thomas Hughes, and Trevor Pinch, eds. 1989. *The social construction of technological systems: New directions in the sociology and history of technology.* Cambridge, MA and London, England: The MIT Press.

Brenner, Robert. 1977. The origins of capitalist development: A critique of Neo-smithian Marxism. *New Left Review* 104: 25–92.

Bunker, Stephen G., and Paul S. Ciccantell. 2005. *Globalization and the race for resources.* Baltimore, MD: Johns Hopkins University Press.

Burkett, Paul. 2013. Lukács on science: A new act in the tragedy. *Historical Materialism* 21(3): 3–15. https://doi.org/10.1163/1569206X-12341313.

Chan, Wing-Cheuk. 2005. On Heidegger's interpretation of Aristotle: A Chinese perspective. *Journal of Chinese philosophy* 32(2): 539-557. https://doi.org/10.1111/j.1540-6253.2005.00320.x.

Clark, Brett, John Bellamy Foster, and Richard York. 2007. The Critique of Intelligent Design: Epicurus, Marx, Darwin, and Freud and the materialist defense of science. *Theory and Society* 36: 515–546. https://doi.org/10.1007/s11186-007-9046-9.

Clausius, Rudolf. 1867. *The mechanical theory of heat with its applications to the steam-engine and to the physical properties of bodies.* London, England: J. van Voorst.

Daly, Herman E. 1996. Beyond growth: The economics of sustainable development. Boston, MA: Beacon Press.

Darwin, Charles. 2011[1859]. *The origin of species.* London, England: Harper.

Debeir, Jean-Claude, Jean-Paul Deléage, and Daniel Hémery. 1991. *In the servitude of power: Energy and civilization through the ages.* London, England and New Jersey: Zed Books.

Daggett, Cara. 2019. *The birth of energy: Fossil fuels, thermodynamics, energy and the politics of work.* Durham, NC: Duke University Press.

Denemark, A. Robert, and Kenneth P. Thomas. 1988. The Brenner-Wallerstein debate. *International Studies Quarterly* 32(1): 47–65. https://doi.org/10.2307/2600412.

Dewey, John. 2008. *The middle works of John Dewey, volume 10, 1899–1924: Journal articles, essays, and miscellany published in the 1916–1917 period.* Carbondale: Southern Illinois University Press.

Dorninger, Christian, and Alf Hornborg. 2015. Can EEMRIO analyses establish the occurrence of ecologically unequal exchange? *Ecological Economics* 119: 414–418. https://doi.org/10.1016/j.ecolecon.2015.08.009.

Dorninger, Christian, Alf Hornborg, David J. Abson, Henrik von Wehrden, Anke Schaffartzik, Stefan Giljum, John-Oliver Engler, Robert L. Feller, Klaus Hubacek, and Hanspeter Wieland. 2021. Global patterns of ecologically unequal exchange: Implications for sustainability in the 21st century. *Ecological Economics* 179: 106824. https://doi.org/10.1016/j.ecolecon.2020.106824.

Drengson, Alan R. 1995. *The practice of technology: Exploring technology, ecophilosophy, and spiritual disciplines for vital links.* Albany, NY: SUNY Press.

Dusek, Val. 2006. *Philosophy of technology: An introduction.* Oxford, England: Blackwell.

Eiseley, Loren. 1969. *The unexpected universe.* San Diego, CA: Harcourt, Brace and World.

Ellul, Jacques. 1964. *The technological society.* New York, NY: Vintage.

Engels, Frederick. 1964[1925]. *Dialectics of nature.* 3rd ed. Moscow, Russia: Progress Publishers.

Engels, Frederick. 1972[1874]. On authority. In *Marx-Engels reader,* edited by Robert C. Tucker, 730–733. 2nd ed. New York, NY: W.W. Norton and Co.

Feenberg, Andrew. 1991. *Critical theory of technology.* New York: Oxford University Press.

Feenberg, Andrew. 2008. Critical theory of technology: An overview. In *Information technology in librarianship: New critical approaches,* edited by Gloria Leckie, and John Buschmanm, 31–46. Westport, CT: Libraries Unlimited.

Feenberg, Andrew. 2009. What is philosophy of technology? In *International handbook of research and development in technology education*, edited by Alister Jones, and Marc J. de Vries, 159–166. Rotterdam, the Netherlands: Sense.

Feenberg, Andrew. 2014. *The philosophy of praxis: Marx, Lukács and the Frankfurt school*. London, England and New York, NY: Verso.

Foster, John Bellamy. 2000. *Marx's ecology: Materialism and nature*. New York, NY: Monthly Review Press.

Georgescu-Roegen, Nicholas. 1971. *The entropy law and the economic process*. London, England: Oxford University Press.

Georgescu-Roegen, Nicholas. 1975. Energy and economic myths. *Southern Economic Journal* 41(3): 347–381. https://doi.org/10.2307/1056148.

Giljum, Stefan, and Nina Eisenmenger. 2004. North-South trade and the distribution of environmental goods and burdens: A biophysical perspective. *Journal of Environment and Development* 13(1): 73–100. https://doi.org/10.1177/1070496503260974.

Haraway, Donna. 1991. *Simians, cyborgs and women: The reinvention of nature*. New York, NY: Routledge.

Harris, Marvin. 1980. *Cultural materialism: The struggle for a science of culture*. New York, NY: Vintage.

Headrick, Daniel R. 2010. *Power over peoples: Technology, environments, and Western imperialism, 1400 to the present*. Princeton and Oxford: Princeton University Press.

Heidegger, Martin. 1977[1954]. *The question concerning technology and other essays*. New York, NY and London, England: Garland Publishing.

Heikkurinen, Pasi. 2018. Degrowth by means of technology? A treatise for an ethos of releasement. *Journal of Cleaner Production* 197: 1654–1665. https://doi.org/10.1016/j.jclepro.2016.07.070.

Heikkurinen, Pasi, Jenny Rinkinen, Timo Järvensivu, Kristoffer Wilén, and Toni Ruuska. 2016. Organising in the Anthropocene: An ontological outline for ecocentric theorising. *Journal of Cleaner Production* 113: 705–714. https://doi.org/10.1016/j.jclepro.2015.12.016.

Helmholtz, Hermann von. 2001[1862]. On the conservation of force: Introduction to a series of lectures delivered in the winter of 1862–1863. In *The Harvard classics: Scientific papers*, edited by Charles Eliot. New York, NY: P.F. Collier and Son. pp. 173–210.

Hickel, Jason, and Giorgos Kallis. 2019. Is green growth possible? *New Political Economy* 25(4): 1–18. https://doi.org/10.1080/13563467.2019.1598964.

Hickman, Larry. 2014. Putting pragmatism (especially Dewey's) to work. In *Philosophy of technology: The technological condition*, edited by Scharff, C. Robert, and Val Dusek, 406–425. 2nd ed. Chichester, West Sussex: Blackwell.

Hornborg, Alf. 1992. Machine fetishism, value, and the image of unlimited good: Toward a thermodynamics of imperialism. *Man* 27(1): 1–18. https://doi.org/10.2307/2803592.

Hornborg, Alf. 1998. Towards an ecological theory of unequal exchange: Articulating world systems theory and ecological economics. *Ecological Economics* 25(1): 127–136. https://doi.org/10.1016/S0921-8009(97)00100-6.

Hornborg, Alf. 2001. *The power of the machine: Global inequalities of economy, technology, and environment*. Walnut Creek, CA: Altamira Press.

Hornborg, Alf. 2003. The unequal exchange of time and space: Towards a non-normative ecological theory of exploitation. *Journal of Ecological Anthropology* 7(1): 4–10.

Hornborg, Alf. 2006. Footprints in the cotton fields: The Industrial Revolution as time-space appropriation and environmental load displacement. *Ecological Economics* 59(1): 74–81. https://doi.org/10.1016/j.ecolecon.2005.10.009.

Hornborg, Alf. 2009. Zero-sum world: Challenges in conceptualizing environmental load displacement and ecologically unequal exchange in the world-system. *International Journal of Comparative Sociology* 50: 237–262. https://doi.org/10.1177/0020715209105141.

Hornborg, Alf. 2014. Why solar panels don't grow on trees: Technological utopianism and the uneasy relation between Marxism and Ecological Economics. In *Green utopianism: Perspectives, politics and micro-practices*, edited by Johan Hedrén, and Karin Bradley, 76–97. New York, NY: Routledge.

Hornborg, Alf. 2016. *Global magic: Technologies of appropriation from ancient Rome to Wall Street*. London, England: Palgrave Macmillan.

Hornborg, Alf. 2017. Artifacts have consequences, not agency: Toward a critical theory of global environmental history. *European Journal of Social Theory* 20(1): 95–110. https://doi.org/10.1177/1368431016640536.

Hornborg, Alf, J. R. McNeill, and Joan Martinez-Alier, eds. 2007. *Rethinking environmental history: World-system history and global environmental change*. Lanham, MD and New York, NY: Altamira Press.

Huesemann, Michael, and Joyce Huesemann. 2011. *Techno-fix: Why technology won't save us or the environment*. Gabriola, BC, Canada: New Society.

Illich, Ivan. 1973. *Tools for conviviality*. Glasgow, Scotland: Collins.

Jorgenson, Andrew K., Kelly Austin, and Christopher Dick. 2009. Ecologically unequal exchange and the resource consumption/environmental degradation paradox: A panel study of less-developed countries, 1970–2000. *International Journal of Comparative Sociology* 50(3–4): 236–284. https://doi.org/10.1177/0020715209105142.

Kline, Ronald. 1995. Construing 'Technology' as 'Applied Science': Public rhetoric of scientists and engineers in the United States, 1880–1945. *Isis: A Journal of the History of Science* 86: 194–221.

Lange, Frederick A. 1925. *The history of materialism*. New York, NY: Harcourt.

Latour, Bruno. 1988. Mixing humans and nonhumans together: The sociology of a door-closer. *Social Problems* 35(3): 298–310.

Latour, Bruno. 1993. *We have never been modern*. Cambridge, MA: Harvard University Press.

Latour, Bruno. 1996: On actor-network theory: A few clarifications. *Soziale Welt* 47(4): 369–381.

Latour, Bruno. 2004. *Politics of nature: How to bring the sciences into democracy*. Cambridge, MA: Harvard University Press.

Law, John. 1999. After ANT: Complexity, naming and topology. *The Sociological Review* 47(S1): 1–14. https://doi.org/10.1111/j.1467-954X.1999.tb03479.x.

Law, John. 2009. Actor network theory and material semiotics. In *The New Blackwell Companion to Social Theory*, edited by Bryan S. Turner, 141–158. Oxford, England: Blackwell.

Law, John, and Annemarie Mol. 1995. Notes on materiality and sociality. *The Sociological Review* 43(2): 274–294. https://doi.org/10.1111/j.1467-954X.1995.tb00604.x.

Layton, Edwin T. 1974. Technology as knowledge. *Technology and Culture* 15(1): 31–41. https://doi.org/10.2307/3102759.

Likavčan, Lukáš, and Manuel Scholz-Wäckerle. 2018. Technology appropriation in a de-growing economy. *Journal of Cleaner Production* 197: 1666–1675. https://doi.org/10.1016/j.jclepro.2016.12.134.

Lotka, Alfred J. 1956. *Elements of mathematical biology*. New York, NY: Dover Publications.

Lucretius, Titus C. 2007. *The nature of things*. London, England: Penguin Books.

Lukács, Georg. 1968[1923]. *History and class consciousness: Studies in Marxist dialectics.* Cambridge, MA: MIT Press.

MacDuffie, Allen. 2014. *Victorian literature, energy, and the ecological imagination.* Cambridge and New York, NY: Cambridge University Press.

MacGregor, Wise. 1991. Intelligent agency. *Cultural Studies* 12(3): 410–428.

Mach, Ernst. 1911. *History and root of the principle of the conservation of energy.* Chicago, IL: The Open Court.

Malm, Andreas. 2016. *Fossil capital: The rise of steam power and the roots of global warming.* London, England: Verso.

Marx, Karl. 1990[1867]. *Capital: A critique of political economy.* London, England: Penguin.

Marx, Karl. 1993[1939]. *Grundrisse: Foundations of the critique of political economy.* London, England: Penguin.

Marx, Karl. 2000[1841]. The difference between the Democritean and Epicurean philosophy of nature. Published 2000. http://marxists.anu.edu.au/archive/marx/works/1841/dr-theses/index.htm. Accessed 11-10-2020.

Marx, Karl. 2000[1844]. Economic and philosophic manuscripts of 1844. Published 2009. https://www.marxists.org/archive/marx/works/1844/manuscripts/preface.htm. Accessed 11-10-2020.

Marx, Leo. 2010. Technology: The emergence of a hazardous concept. *Technology and Culture* 51(3): 561–577.

Merchant, Carolyn. 1980. *The death of nature: Women, ecology, and the scientific revolution.* New York, NY: Harper.

Merchant, Carolyn. 2014. Ecofeminism and feminist theory. In *Environmental ethics,* edited by Michael Boylan. 2nd ed., 59–63. Chichester, West Sussex: Wiley-Blackwell.

Mills, Mary B. 2003. Gender and inequality in the global labor force. *Annual Review of Anthropology* 32: 41–62.

Mitcham, Carl. 1994. *Thinking through technology: The path between engineering and philosophy.* Chicago and London, England: The University of Chicago Press.

Mulvaney, Dustin. 2019. *Solar power: Innovation, sustainability and environmental justice.* Oakland: University of California Press.

Mumford, Lewis. 1954. *Technics and civilization.* London, England: Routledge and Kegan Paul Ltd.

Mumford, Lewis. 1964. Authoritarian and democratic technics. *Technology and Culture* 5(1): 1–8. https://doi.org/10.2307/3101118.

Newell, Peter. 2021. *Power shift: The global political economy of energy transitions.* Cambridge: Cambridge University Press.

Ngram Viewer. 2020a. Technology, technique, machine, 1800–2019, English (2019), case-insensitive, smoothing of 5. *Google Books Ngram Viewer.* Published 05-12-2020. http://books.google.com/ngrams. Accessed 05-12-2020.

Ngram Viewer. 2020b. Technology, technique, machine, 1800–2019, French (2019), case-insensitive, smoothing of 5. Published 05-12-2020. *Google Books Ngram Viewer.* http://books.google.com/ngrams. Accessed 05-12-2020.

Ngram Viewer. 2020c. Technology, technique, machine, 1800–2019, German (2019), case-insensitive, smoothing of 5. Published 05-12-2020. *Google Books Ngram Viewer.* http://books.google.com/ngrams. Accessed 05-12-2020.

OECD. 2011. Towards green growth: A summary for policymakers. Published 25-05-2011. https://www.oecd.org/greengrowth/48012345.pdf. Accessed 20-11-2020.

Oldfield, Jonathan D., and Denis J. B. Shaw. 2013. V.I. Vernadskii and the development of biogeochamical understandings of the biosphere, c. 1880s-1968. *The British Journal for the History of Science* 46(2): 287–310. https://doi.org/10.1017/S0007087412000015.

Pellizzoni, Luigi. 2011. Governing through disorder: Neoliberal environmental governance and social theory. *Global Environmental Change* 21: 795–803. https://doi.org/10.1016/j.gloenvcha.2011.03.014.

Pomeranz, Kenneth. 2000. *The great divergence: China, Europe, and the making of the modern world economy*. Princeton, NJ and Oxford, England: Princeton University Press.

Prieto, Pedro A., and Charles A. S. Hall. 2013. *Spain's photovoltaic revolution: The energy return on investment*. New York, NY: Springer.

Rabinbach, Anson. 1992. *The human motor: Energy, fatigue, and the origins of modernity*. Berkley and Los Angeles: University of California Press.

Rappaport, Roy. 1968. *Pigs for the ancestors: Ritual in ecology of a New Guinea people*. New Haven, CT and London, England: Yale University Press.

Rice, James. 2007. Ecological unequal exchange: Consumption, equity, and unsustainable structural relationships within the global economy. *International Journal of Comparative Sociology* 48(1): 43–72. https://doi.org/10.1177/0020715207072159.

Roos, Andreas. 2021. Earthing philosophy of technology: A case for ontological materialism. In *Sustainability beyond technology: Philosophy, critique, and implications for human organization*, edited by Pasi Heikkurinen, and Toni Ruuska, 58–98. Oxford: Oxford University Press.

Ruuska, Toni, Pasi Heikkurinen, and Kristoffer Wilén. 2020. Domination, power, supremacy: Confronting anthropolitics with ecological realism. *Sustainability* 12(7): 2617. https://doi.org/10.3390/su12072617.

Scharff, C. Robert, and Val Dusek, eds. 2014. *Philosophy of technology: The technological condition*. 2nd ed. Chichester, West Sussex: Wiley-Blackwell.

Schmidt, Alfred. 2014[1971]. *The concept of nature in Marx*. London, England and New York, NY: Verso.

Schrödinger, Erwin. 1945. *What is life? The physical aspect of the living cell*. London, England: Cambridge University Press.

Seamster, Louise, and Victor Ray. 2018. Against teleology in the study of race: Toward the abolition of the progress paradigm. *Sociological Theory* 36(4): 315–342. https://doi.org/10.1177/0735275118813614.

Soper, Kate. 1995. *What is nature?* Oxford, England: Wiley-Blackwell.

Spencer, Herbert. 1904. *Essays, scientific, political, and speculative*. New York: D. Appleton and Company.Stanford, Kyle P. 2006. Instrumentalism: Global, local, and scientific. In *The Oxford handbook of philosophy of science*, edited by Paul Humphreys, 318–336. New York, NY: Oxford University Press.

Steffen, Will, Jacques Grindevald, Paul Crutzen, and John McNeill. 2011. The Anthropocene: Conceptual and historical perspectives. *Philosophical Transactions of the Royal Society* 369: 842–867. https://doi.org/10.1098/rsta.2010.0327.

Stock, Ryan, and Trevor Birkenholtz. 2019. Photons vs. firewood: Female (dis)empowerment by solar power in India. *Gender, Place and Culture* 27(1): 1628–1651. https://doi.org/10.1080/0966369X.2020.1811208.

Stokes, Kenneth M. 1994. *Man and the biosphere: Toward a coevolutionary political economy*. Armonk, NY and London, England: M.E. Sharpe.

Suess, Eduard. 1875. *Die Entstehung der Alpen*. Vienna, Austria: W. Braumüller.

UNEP. 2014. *Decoupling 2: Technologies, Opportunities and Policy Options. A Report of the Working Group on Decoupling to the International Resource Panel.* Nairobi, Kenya: UNEP.

Vernadsky, Vladimir I. 1945. The biosphere and the noösphere. *American Scientist* 33(1): 1–12.

Vernadsky, Vladimir I. 1998[1929]. *The biosphere.* New York, NY: Springer-Verlag.

Vernadsky, Vladimir I. 2001[1938]. Problems of biogeochemistry II: On the fundamental material-energetic distinction between living and nonliving natural bodies of the biosphere. *21st Century Science and Technology* 13(4): 20–39.

Vogel, Steven. 1996. *Against nature: The concept of nature in critical theory.* Albany: State University of New York Press.

Wallerstein, Immanuel. 2004. *World-systems analysis: An introduction.* Durham, NC and London, England: Duke University Press.

Wallerstein, Immanuel. 2011a. *The modern world system I: Capitalist agriculture and the origins of the European world-economy in the sixteenth century.* Berkeley, Los Angeles and London, England: University of California Press.

Wallerstein, Immanuel. 2011b. *The modern world system II: Mercantilism and the consolidation of the European world-economy, 1600–1750.* Berkeley, Los Angeles and London, England: University of California Press.

Wallerstein, Immanuel. 2011c. *The modern world system III: The second era of great expansion of the capitalist world-economy, 1730–1840.* Berkeley, Los Angeles and London, England: University of California Press.

Wallerstein, Immanuel. 2011d. *The modern world system IV: Centrist liberalism triumphant, 1789–1914.* Berkeley, Los Angeles and London, England: University of California Press.

Warlenius, Rikard. 2016. Linking ecological debt and ecologically unequal exchange: Stocks, flows, and unequal sink appropriation. *Journal of Political Ecology* 23(1): 364–380. https://doi.org/10.2458/v23i1.20223.

Weart, Spencer. 2003. *The discovery of global warming.* Cambridge, MA and London, England: Harvard University Press.

Wilding, Adrian. 2010. Naturphilosophie redivivus: On Bruno Latour's 'political ecology.' *Cosmos and History: The Journal of Natural and Social Philosophy* 6(1): 18–32.

Winner, Langdon. 1979. The political philosophy of alternative technology: Historical roots and present prospects. *Technology in Society* 1(1): 75–86. https://doi.org/10.1016/0160-791X(79)90010-1.

Winner, Langdon. 1980. Do artifacts have politics? *Daedalus* 109(1): 121–136.

Winner, Langdon. 2020[1986]. *The whale and the reactor: A search for limits in an age of high technology.* Chicago and London, England: The University of Chicago Press.

WWF. 2020. *Living planet report 2020: Bending the curve of biodiversity loss.* Gland, Switzerland: WWF.

Yang, Dennis T., Vivian W. Chen, and Ryan Monarch. 2010. Rising wages: Has china lost its global labor advantage? *Pacific Economic Review* 15(4): 482–504. https://doi.org/10.1111/j.1468-0106.2009.00465.x.

3 The historical context of solar technology

As in the case of Katrineholm, conventional solar visions are often rooted in an understanding of technological change as occurring in isolation from the international exchange of resources in the world economy. This chapter sketches out an interdisciplinary understanding of the most recent major energy transformation – the Industrial Revolution – in order to create a more realistic understanding of technological change. The broader goal is to understand the ramifications of altering the metabolic basis of industrial societies by means of solar photovoltaic (PV) technology. I explain and draw from the interdisciplinary concept of "socio-metabolic regimes" to understand these ramifications. This is followed by an analysis of how the Industrial Revolution both necessitated and facilitated specific ecological conditions, social relations of power, and cultural imaginaries, which are not easily separated from today's development solar PV technology. The chapter then identifies three different interpretations of the Industrial Revolution, which all disagree on the prospects of leaving the industrial regime by means of renewable energy technologies. I show that the answer to whether solar PV technology is based on ecologically unequal exchange may inform us whether solar PV technology is (i) a means to continue the metabolic basis of industrial societies, (ii) a means to transcend it, or (iii) a means to reverse it.

What is a metabolic regime?

Numerous frameworks have been developed by social scientists to understand how energy technologies at once arise from and generate particular social relations (Markard et al., 2012; Geels, 2012, 2014; Sovacool, 2016). However, as we saw in the previous chapter, the most relevant approaches to energy technology are those that acknowledge how energy transformations are biophysical events with relevance in the natural world. By drawing on new sources of material and energy, energy transformations are bound to alterations in the natural environment and of nature's processes. As recently suggested by environmental historians Ian Jared Miller et al., "energy transitions are always at some level also environmental events" (2019: 466). This means that the unit of analysis must include not only the socio-technical formation associated with a particular energy regime, but the socio-technical formation *within its environment*. Many of the interdisciplinary

DOI: 10.4324/9781003292319-3

studies that have been conducted in this vein have concluded that to leave the fossil regime implies a transformation that is comparable to the Neolithic and Industrial Revolutions (Debeir et al., 1991; Podobnik, 2005; Fischer-Kowalski and Haberl, 2007; Lenton et al., 2016).

This means that the history of human-environmental relations is not the history of humans or the history of environments, but a history of the relation between the two. Weisz et al. (2001: 122) summon Maurice Godelier to argue that humans "transform their relation *with* nature by transforming *nature itself*."[1] In turn, these transformations, which affect the dynamic biogeochemical processes of nature itself, provoke the historical opportunity or necessity for social-technical changes. This social-ecological dialectic is the underlying methodological departure for the study of a range of phenomena, such as "energy systems," "dialectical materialism," "raw materialism," "ecological-economic history," and "ecologically unequal exchange." We can say that these approaches all in one way or another deal with the process of "coevolution" wherein societies are "structurally coupled to parts of their environment, leading to a process where both mutually constrain each other's future evolutionary options" (Weisz et al., 2001: 123; Kallis and Norgaard, 2010).

The concept of "social-ecological regimes" (Fischer-Kowalski and Haberl, 2007; Krausmann et al., 2016) provides a helpful systematization of human-environmental coevolution in history. Essentially, a social-ecological regime denotes "a specific fundamental [metabolic] pattern of interaction between (human) society and natural systems" (Fischer-Kowalski and Haberl, 2007: 8). These fundamental patterns may contain different forms of social relations and are therefore broader historical categories than Marxist "modes of production," which are defined foremost by historically developed social relations (Debeir et al., 1991: 12).[2] The essential difference between different social-ecological regimes is how human populations harness matter-energy from their surrounding environment. Crucially, however, such metabolic interactions cannot fully be understood without considering how internal social relations facilitate and reproduce them. Social interaction, its preconditions and form, alters continually in relation to the dynamic changes of natural environments. In the study of social-ecological regimes, culture, understood as the "the total socially acquired life-style of a group of people including patterned, repetitive ways of thinking, feeling and acting," is similarly understood as facilitating the reproduction of a given social metabolism (Harris, 1997: 88; Weisz et al., 2011). Alongside environmental changes, transitions from one socio-ecological regime to another are therefore associated with major historical alterations in both social relations and cultural imaginaries.[3]

The rise of the industrial regime

There is no doubt that the Industrial Revolution marks the beginning of historically unprecedented changes in human-environmental relations. Much of the recent work in environmental history considers the Industrial Revolution as the analytical focal point *par excellence* for understanding the roots of the global environmental problems of today (Barca, 2011; Steffen et al., 2015a, b; Malm, 2016a, b). However, as these studies show, more than fundamentally altering the environment on a global scale, the Industrial Revolution marked the beginning of an entirely new form of social-ecological regime. This is an important point, because it is crucial to acknowledge how today's global environmental challenges emerged in tandem with specific social relations and cultural ideas that took shape with the historically new means to harness highly energy-dense fossil energy.

As in the other two major energy transformations in history, the taming of fire and the Neolithic Revolution, it is difficult to isolate the exact causality behind the emergence of the industrial social-ecological regime in a satisfactory way (Weisdorf, 2005). This is not to say that scholars have not attempted the task of doing so. To date, causal explanations range from the classical notion of human (notably British) engineering ingenuity in the design of the steam engine following the Enlightenment, to the British factory owners' desire to dominate labor in the process of accumulating more capital. Western scholars have typically thought of the Industrial Revolution as something emerging first in Britain around 1760, as an effect of some cultural, social, or ecological factor internal to Britain. This approach still dominates today, even if numerous historians have made much effort to demonstrate the flaws in such reductive and Eurocentric approaches (Frank, 1966; Wolf, 2010[1982]; Pomeranz, 2000; Inikori, 2002; Marks, 2015). Acknowledging that the emergence of industrialism in Britain was contingent upon international relations – as exemplified by the increasing role of the international market, the slave trade, and raw material imports from colonies – implies in turn that any attempt to understand the birth of the industrial regime must be attentive to historical patterns established well before the year 1760. These include understanding capitalism and colonialism as a world strategy for ecological appropriation, emerging already in the late 15th century. Here, we shall try to understand colonialism and capitalism as two interconnected prerequisites for the emergence of the industrial regime. In addition to these prerequisites, it is widely acknowledged that the industrial regime could not emerge without burning fossil fuels to propel machines used in the mass production of commodities for the world market.

Colonialism and capitalism as prerequisites for the Industrial Revolution

If industrialism refers to "an inclination toward mass production of commodities," then we have good reason to think that the industrial regime could not emerge without an already significant level of social-political complexity (Hornborg, 2015: 863). This is so because mass production cannot exist without the mass extraction of matter-energy inputs and a large-scale system of labor orchestration. For the ecological economist Robert Ayres (2016: 389), "[e]nergy (exergy) availability was the main engine of growth from the start of the Industrial Revolution." Without access to copious amounts of biomass from the world peripheries, it would have been impossible to extract and burn fossil fuels.[4] From this, we can deduce that the industrial regime must have emerged from a "successful" agrarian social me-tabolism that drew resources from an immense land base. As some environmental historians have shown, such complexity was generated through colonial expansion of European imperial powers, which appropriated natural resources and displaced entropy around the globe through means of trade, warfare, and slavery (Crosby, 1986; Pomeranz, 2000; Moore, 2003; Hornborg, 2006).

Environmental historian Jason Moore (2003, 2007) has thoroughly examined the mechanisms and root causes of the capitalist world system and its relation to early European colonialism. Drawing on the work of Immanuel Wallerstein, Moore provides an interdisciplinary interpretation of the rise of capitalism that considers "nature and society as mutually relational" (2003: 357). In seeking a "synthesis of theory and history for the study of large-scale socio-ecological change over the *long duree*," Moore's early work on the origins of capitalism is an indispensable contribution to the environmental history of social-ecological regimes (ibid. 308). According to Wallerstein and Moore, the capitalist world economy emerged out of the social-ecological crisis in 15th-century European feudalism that had "overstepped the socio-ecological limits to continued expansion" (Moore, 2003: 313, see also Wilkinson, 1973). For Wallerstein (2011: 37), the most plausible explanation for this crisis was that "[a]fter a thousand years of surplus appropriation under the feudal mode, a point of diminishing returns had been reached." Feudalism, in the process of producing a surplus for the ruling strata, tended not to reinvest in the soil, which eventually led to soil exhaustion and diminishing returns (Moore, 2003: 330; Wallerstein, 2011: 23).

In the crisis of feudalism, ecological factors such as climatological change and the Black Death were important catalyzers for the emergence of capitalism. The Black Death, in particular, was a significant event because it fundamentally altered the power structure of feudal Europe through changes in demography. The effects in Europe were devastating, leading to the death of 40%–60% of the population. In turn, this had serious repercussions for the production of surplus for the managerial elites, who found themselves without the provisions generated through serf labor. The dissolution of "the feudal equivalent of the 'reserve army of labor'" implied that the peasants bargaining position drastically improved (Moore, 2003: 314). To put it crudely, the "supply" of peasant serfs drastically diminished, thereby increasing their perceived value. The result was that "Europe's peasantry waged the class struggle much more effectively than heretofore, squeezing the seigneurs, who in turn squeezed the states, who were forced to recognize the former's voice in policy-making" (ibid. 2003: 317). Since the European peasantry resisted increased exploitation, often in outright rebellion, "Europe's ruling strata" came to favor "'outer' rather than 'inner' expansion" to satisfy their need for surplus and political power (Wallerstein, 2011; quoted in Moore, 2003: 316).

Internal reforms, such as converting arable land to pasturage in Castile and England, were attempted in the hope of increasing the production of surplus value by producing wool for export. In the end, these were not very effective and "transatlantic expansion was the path of least resistance" (Moore, 2003: 316). Meanwhile, stories of a New World in the west were spreading. Christopher

Columbus, in a speech to the Court in Madrid, gave an alluring account of his travels to what he believed to be a group of islands just east of Asia:

> Hispaniola is a miracle. Mountain and hills, plains and pastures, are both fertile and beautiful … the harbors are unbelievably good and there are many wide rivers of which the majority contain gold … There are many spices, and great mines of gold and other metals (Zinn, 1995: 3).

Columbus's choice of words in describing Hispaniola (known by the indigenous peoples as Haiti or Quizqueia) was not accidental. Nor was it romantic. It was a description of a financial opportunity for commerce and surplus production. Fertile lands, good harbors and wide rivers were all optimal conditions for resource exploitation and extraction. Precious metals such as gold and silver were extremely desirable goods on the emerging world market and were therefore likely to attract investment opportunities from commercial city-states such as Genoa, who financed the Iberian colonial expansion. Silver, in particular, was highly sought after due to an enormous Chinese demand stemming from Chinese financial reforms (from ca. 1400) that remonetized the economy from paper and copper coinage to silver as a medium of accounting and store of value (Pomeranz, 2000: 159–162).

In Europe, the increased quantities of bullion extracted from the Americas, such as gold and silver, functioned as a foundation for the emerging international market. As grain prices went up in one part of Europe, for example, the bullion functioned as a means for purchasing grain from elsewhere in Europe, thereby stabilizing prices in a common market sphere. As such, Wallerstein argues, increased access to bullion operated as a "hedge" and "sustained the thrust of the expansion, protecting [the] still weak system against the assaults of nature" (2011: 45–46, 76). The result was the emergence of a more resilient form of world commerce operating with a world division of labor under the logic of commodification, i.e., the capitalist system (ibid. 77).

World trade in luxury goods, or "preciosities," such as silk, carpets, and spices had existed for some time, but it had done little to alter or affect regional metabolisms in Europe to the degree that it upset domestic power relations (Wallerstein, 2011: 41–42). In China, in contrast, the intensified production of silk (among other goods) for international trade contributed to considerably changing social and ecological conditions (Abu-Lughod, 1989). With the expanding capitalist market, trade in staples such as wheat, wood, and sugar proved to have tangible material effects both in the regions that benefited from the trade and in those regions that carried the consequences. Wheat and wood imported from the Baltic (part of Europe's "internal Americas") freed up land in core regions that could be converted to pastures for draft animals and livestock, while simultaneously feeding a growing labor force and serving as raw material for house- and shipbuilding. Sugar, an emerging food staple in the core, was first planted by the Portuguese with slave labor on the island of Madeira, and this model was exported to the Americas where it was expanded and intensified to provide increased revenues for the newly formed coalitions between the merchant classes, the bankers, and the ruling classes of Europe. Sugar, in particular, was dependent upon slave labor and both the colonial production of sugar and silver (and later cotton and tobacco) were extremely destructive and inhumane processes. The detrimental effects of sugar plantations led to what Moore (2003) calls "sequential exploitation," whereby the ruthless treatment of landscapes and slaves through extractive mono-cropping forced colonial plantation capitalists to move from region to region throughout the Americas in order to maintain profitable yields.

From a wider, world-economic point of view, this degradation occurred for the benefit of those who could enjoy the final products. Early capitalist expansion and colonialism was ultimately a process whereby "[n]utrients [and all manner of resources] flowed from country to city in the New World, and thence from urban centers in the periphery to the core" (Moore, 2003: 334; see also Hornborg, 2006; Clark and Foster, 2009). The appropriation of resources and raw materials from the world peripheries contributed to an accumulation of capital in the world core, spurring the growth of a modern middle class, increased technological development and the further manufacture of cheap commodities for the world market (Hornborg, 2006). The result was not only detrimental social and ecological effects in the peripheries, but also that some European nations, such as Britain, could uphold a metabolic throughput larger than the biocapacity of their domestic land base (Pomeranz, 2000).

Even in colonial centers, however, the material bounty collected from around the world was unequally distributed. In core regions of the world, such as Britain, the "country to city" flow of nutrients was a central aspect of the emergence of capitalist class relations, ultimately contingent upon the "enclosure acts" by which landowners displaced peasants from their land to be used as sheep pasture (Foster, 2000: 172; Hornborg, 2015). The displaced peoples congregated in the cities where they made up the new class of the proletariat, a "reserve army of labor" condemned as serfs for capitalist industrial manufacturing (Marx, 1990[1867]). The new capitalist class gaining influence through the process of commodity production eventually grew powerful enough to challenge the tributary-based state system and so the power structure was recomposed (Wolf, 2010[1982]: 265–266).

In this way, the rise of capitalism was inseparable from European colonialism and world commerce. Since its inception, then, capitalism has been contingent upon a world division of labor upholding an unequal distribution of social-ecological benefits and burdens (Moore, 2003, 2007; Wallerstein, 2011; Hornborg, 2013). In effect, "where earlier ecological crises had been local, capitalism globalized them. And it did so at a pace that outstripped all previous existing historical systems" (Moore, 2003: 303; see also Clark and Foster, 2009). Capitalism, in this sense, is inherently global in character, defined by the world social relations that allowed for the rise and continuation of resource accumulation among European imperial powers. Compared to previous empires, capitalism could (and did) draw on resources from around the globe (Wolf, 2010[1982]). According to Moore (2003), this distinctive feature was the foundation for a completely new form of world economy in which the European capitalist resource base greatly exceeded its predecessors both in absolute and relative terms. This was, in other words, a historically unique social condition distinct from feudalism. In Marx's words (1990[1867]: 915):

> The discovery of gold and silver in America, the extirpation, enslavement and entombment in mines of the indigenous population of that continent, the beginnings of the conquest and plunder of India, and the conversion of Africa into a preserve for the commercial hunting of blackskins, are all things which characterize the dawn of the era of capitalist production. These idyllic proceedings are the chief moments of primitive accumulation.

The term "primitive accumulation" is used here to denote the "prehistory and the precondition of capital" (Foster, 2000: 173, emphasis added). It was only within a highly exploitative world economy that the resources necessary for industrialization could be acquired, maintained, and increased.

Fossil fuels as a prerequisite for the industrial regime

Even if industrial mass production was first propelled by hydropower, fossil energy played a definite role for the industrial regime, such as it developed in the 19th and 20th centuries (Malm, 2016a, b). Beyond the role of coal as a source of energy in 19th-century steam-powered manufacturing, fossil gas and oil were later pivotal for the development of industrial agriculture based on artificial fertilizer and for a range of artificial material compounds that replaced naturally occurring materials (Smil, 2001, 2016b).

To understand the importance of fossil energy for the historical development of the industrial regime, we need first an informed understanding of what fossil energy is. This begins with an acknowledgment that fossil fuels originate as organic material that has been transformed over millions of years under conditions of enormous pressure and high heat in the crust of the Earth. This organic material was once the bustling ecosystems of prehistoric plants and animals that captured direct energy from the sun through photosynthesis and built themselves from material found in the air, water, and soil. Over several hundred million years, layer upon layer of dead organic material accumulated in anaerobic environments (such as swamps). These layers were then buried under each other and pushed down in the Earth's crust where they were subjected to extreme heat and pressure that eventually transformed them into the coal, oil, or fossil gas that industrialists extract today. As such, fossil fuels represent millions of years' worth of highly energy-dense "buried sunshine" (Dukes, 2003).

The fact that modern humans orchestrate collective efforts to release the energetic potential of this fossilized solar energy in – by geological standards – the blink of an eye, helps to explain the unprecedented historical condition of the industrial regime. Swedish journalist Therese Uddenfeldt pedagogically appeals to our sensory experiences as she explains the fundamental difference between direct solar energy and fossil energy. She writes (2016: 20, my translation),

A simple experiment: First, reach out with one of your hands into a beam of sunshine and enjoy. Then put a piece of burning coal in the other. In the first hand, you experience the effects of direct sunshine. In the second hand, concentrated sunshine is in the process of melting your skin. What you experience is the difference between how much direct sunshine and concentrated sunshine can accomplish.

Note that the difference between the two energy sources is not a difference in original energy source, i.e., sunshine. Coal, however, is a concentrated form of sunshine that is highly energy dense. The high energy density in the coal is contingent upon the time it took for it to accumulate and compress in the Earth's crust, and the space for the original sunshine to hit the Earth and to be sequestered by ancient photosynthesizing plants. The sheer scales and time-spans in question make it impossible for humans to imitate the formation of fossil fuels without great energetic costs (see, e.g., Nikiforuk, 2012: 96). It is, in other words, energetically futile to attempt to produce artificial fossil energy carriers in the hopes that these could maintain any human social metabolism. This is why fossil fuels are considered non-renewable, as it simply takes too long for them to be renewed through the biogeochemical processes described above. Access to fossil fuels and the great energetic potential they embody is by all practical accounts a historically one-time opportunity (for a discussion, see Love and Isenhour, 2016).

The non-renewability of fossil energy carriers does not preclude the fact that they are highly lucrative sources and substitutes for both space and time (Hornborg et al., 2019). Space and time, both represented in fossil energy, correspond to the two categories of land and labor that had a primary significance for the expansion and increased complexification of social metabolisms in the agrarian regime (Sieferle, 2001). Let us briefly look at these two categories in turn.

Fossil energy as compressed space. In Uddenfeldt's "simple experiment," it becomes easier to understand how the transition from the reliance upon direct sunshine to the reliance upon coal altered the relation to land with the industrial regime. Simply put, the lump of coal held in the second hand can be understood to represent numerous concentrated hands of the first kind. In the terminology of ecological economics, the coal is said to contain huge amounts of "embodied land," i.e., the energetic potential of land areas. As the reliance upon coal grew (first in Britain), the social metabolism increasingly came to rely upon the land embodied in the fossil fuels extracted from mines. The historian Edward Wrigley (1988) famously thought of the British industrialization as a transition from an "organic economy" to a "mineral-based economy" that burst the solar income constraint of the agrarian regime limited by access to land (for a critique, see Malm, 2016a, b). More recent studies have come to similar conclusions of how fossil fuels represent access to space (Bridge, 2010; Huber and McCarthy, 2017). Geographer Gavin Bridge (2010) considers the geography of the industrial regime as a "geography of holes," with reference to how extractive sites of coal, oil, and fossil gas are comparatively small in relation to the amount of energy potential that is extracted.

The concept of "power density," as developed by Vaclav Smil (2015), also captures the relation between land and fossil fuels. In short, power density is a measure of how much space is needed for any given means of capturing energy (W/m^2). The concept is primarily used for understanding the challenges in transitioning away from fossil fuels, but it is equally useful for understanding what it implied to transition to fossil fuels in the first place. For example, Smil (2015: 3–4) has calculated that harvesting oak or beech has a power density of 0.22 W/m^2, while a British coal mine in the 1770s produced fuel with a power density of nearly 1,200 W/m^2.[5] Again, this shows that the industrial regime fundamentally altered the human relation to land and the environment. Given this situation, it becomes evident that there is a fundamental difference between the agrarian regime and the industrial regime regarding the access to the energy necessary for the metabolic process. Whereas the agrarian regime is highly dependent on access to quantities of (preferably fertile) land for the production of consumable biomass, the industrial regime is less dependent upon quantities of land for access to energy, but increasingly reliant upon what we may call "qualitative" lands or "holes" rich in stocks of fossil fuels.

The consequence of the very high energy density of coal was that it allowed matter-energy throughputs to be increased beyond the biocapacity of local environments, including the flow of energy harnessed through waterpower or wind power. Even if fossil energy did not cause the Industrial Revolution, access to fossil energy meant that the "social metabolism [could] be greatly increased without decimating the base of human nutrition" (Weisz et al., 2001: 129).[6]

Fossil energy as compressed time. Apart from embodied land, fossil fuels represent also high concentrations of embodied time. The very essence of industrialism as an inclination toward the mass production of commodities under the capitalist logic of accumulation was founded upon this new energy-time relation. Andrew Nikiforuk (2012: 26) uncovers how early industrial capitalists marveled at the capacity for work embodied in coal as "two pounds of coal could … lift a man to the summit of Mont Blanc without any human toil." The time embodied in coal was used to speed up the manufacturing process and reduce the time for traveling and transportation of goods and materials (Hagens, 2020). For example, prior to the industrialization of the manufacturing process it took up to 50,000 human labor hours to spin 50 kg of cotton. With the fossilization of the cotton industries, the process of spinning the same amount of cotton took only 135 hours (Hornborg, 2010: 122).

Coal, however, could not be fed to human or animal bodies in the hope of converting its potential energy into useful work. These "prime movers"[7] were contingent upon biomass (food, such as grain) to do work. To convert fossil energy, another category of prime mover was necessary. The steam engine, built for this purpose, was first developed by the Spanish mining administrator Jerónimo de

Ayanz, who employed it to pump water out of the silver mine in northern Seville, Spain in 1606. It was in Britain, however, that the steam engine made its true commercial breakthrough (Headrick, 2009: 91–110). The eventual dependency on coal in industrial manufacturing was coupled with a combined dependence on steam engines (Smil, 2016b). One could not exist without the other. In the transition from the agrarian to the industrial regime, to put it succinctly, coal was to machinery what biomass was to animals (Hornborg, 2014). The increased reliance upon the coal-machinery combination was biophysically favorable given its massive energetic potential, but also because it meant that human and animal nutrition did not compete directly with the access to energy for other purposes, such as manufacturing and transportation (Weisz et al., 2001).

The acceleration of the industrial regime

As we have seen, colonialism, capitalism, and use of fossil fuels generated the conditions for a rapidly expanding social metabolism. The consequences of this were nothing short of a revolution with far-reaching repercussions in many aspects of human (and non-human) life. Let us now turn to some of the arguably most important consequences.

Fossil dependency as a consequence of the Industrial Revolution

In *Fossil Capital*, Malm (2016a, b: 288) concludes that at a certain stage in the development of capital, fossil fuels became an indispensable part of capitalist production of surplus value *"across the [entire] spectrum of commodity production,"* in which each round of appropriation generating money was coupled to an increased burning of sources of fossil energy. While most scholars have been content with pointing out how fossil energy facilitated a transgression of the local solar income constraint of the agrarian regime, Malm argues that coal was favored by British cotton manufacturers because it could facilitate capitalist relations of production in which the workers were separated from the means of production, thereby lowering the cost of labor and the price of the British cotton textiles.

The first obvious problem with fossil dependency is that fossil energy carriers are non-renewable. Therefore, by all practical accounts, extracting and burning fossilized sources of energy is a one-time historical opportunity that is subjected to diminishing energy returns. Economist Stanley Jevons was early to point this out in relation to the importance of coal to the continued expansion of the British Empire. In *The Coal Question*, Jevons lamented over how digging deeper would only cause the price of British coal to increase, something that would in turn raise the prices of domestically manufactured goods on the world market and undermine the competitiveness of British manufacture. Import of coal from the colonies, Jevons continued, would also raise prices of British manufactures and so he concluded that the British Empire was doomed to end with the diminishing returns on coal. In this, Jevons was early to uncover the general pattern of industrial metabolism operating with diminishing energy returns on energy invested (EROI) (Jevons, 1865; today, see Hall and Klitgaard, 2012; Hall et al., 2014).

Since fossil energy carriers are non-renewable, depending on them metabolically is inseparable from an imperative to expand the depth and reach of the social metabolism. Typically, the most easily accessible and lucrative sources of fossil energy carriers are extracted first, which means that it becomes increasingly expensive to continue the reliance upon a specific fossil fuel (Hall and Klitgaard, 2012). Relying upon coal as a non-renewable energy carrier thereby incentivized British capitalists to develop and build networks of railways and trade routes that allowed for the extraction of coal in even the most distant areas throughout the world (Debeir et al., 1991; Malm, 2016b).

Given the fact that infrastructure and old extractive equipment quickly become obsolete or inadequate in the face of the recurrent challenges to dig deeper, travel further, or extract faster, industrial capitalism has been forced to constantly renew its means of metabolic throughput. The diminishing EROI of fossilized energy carriers has thereby given rise to a dynamic non-equilibrium in the industrial regime where the social-technical means for appropriation constantly needs to be developed. Alongside "[t]he need of a constantly expanding market for its products," this led "the bourgeoisie over the entire surface of the globe" in search for fossil fuels (Marx and Engels, 1969[1848]; Malm, 2018a). This expanding process is a biophysical imperative in the industrial regime. Essentially, the expansive quest for increased quantities of energy, material resources, and labor for the production of ever-cheaper commodities demands that industrial capitalism enters new areas of nature or society, something that in turn requires a constant development of more advanced technologies. In this sense, "primitive accumulation" is not a historical one-time event, but an ongoing logic of capital expansion that constantly generates new markets and new spheres of appropriation (Luxemburg, 2003[1913]; Harvey, 2003; Bunker, 2007). Considering this imperative, it is legitimate to consider to what degree solar PV technology provides an escape from this extractive logic or whether it is chiefly a way to exploit nature or society further.

To understand this movement in history, Andreas Malm (2018a) historicizes the entry of new fossil energy carriers in relation to what economists call "long waves" of economic development. Throughout the 19th and 20th centuries, in short, the economic surges in the industrial regions of the world have been supported and upheld by either novel ways of extracting and burning fossil energy or upon the discovery and commercialization of entirely new fossil energy carriers (Table 3.1). Complementary figures from Smil (2016a) show that humans and draft animals, propelled by biomass, were the dominant prime movers at the onset of the Industrial Revolution. This, however, changed fast with the coming of the 20th century as the industrial metabolism gained momentum.

These long waves of fossil development are commonly mistaken for energy transformations in two regards. First, it is often assumed that absolute quantities of a previously dominant energy carrier are decreasing when another rises to prominence. The rising prominence of crude oil, for example, is easily mistaken

Table 3.1 Long waves of fossil development in the industrial regime

Economic waves	Upswing	Downswing	Dominant prime mover(s)	New prime movers	Industries and input	Dominant energy carrier(s)	New energy carriers
First wave	1780–1825	1825–1848	Humans and draft animal	Water-powered mechanization, stationary steam engine	Cotton and iron	Biomass	Coal
Second wave	1848–1873	1873–1896	Humans and draft animal, steam engines	Mobile steam engine	Railways, machine-tools, cotton, iron	Biomass	Coal
Third wave	1896–1914	1914–1945	Steam engines	Steam turbines, electric generators	Electrical equipment, engineering, chemicals, steel	Coal	Crude oil, fossil electricity
Fourth wave	1945–1973	1973–1992	Internal combustion engine, electric generators	Nuclear reactors, gas turbines	Automobiles, aircrafts, refineries, petrochemicals	Crude oil, coal, fossil gas	Fossil gas, nuclear electricity
Fifth wave	1992–2008	2008–present	Internal combustion engine, electric generators	Solar PV modules, wind turbines, etc.	Computers, software, microprocessors, fracking	Crude oil, coal, fossil gas	Tar sands, shale oil, renewable electricity

Adapted from Malm (2018a) and revised with reference to Smil (2016a).

as indicating a decrease in the burning of coal. Coal burning is, in fact, increasing even today (Gellert and Ciccantell, 2020). The same applies to the rapid expansion of solar PV technology, which does not prevent the expansion of fossil energy, including crude oil and fossil gas (IEA, 2020). Second, the transition from one type of fossil energy to another (e.g., from coal to oil) is often mistaken for an energy transformation comparable to shifts in socio-metabolic regimes. However, energy sources such as petroleum, fossil gas, tar sands, and shale oil are all non-renewable sources that support a qualitatively similar human-environmental relation. These energy carriers are all made from "buried sunshine" and contingent upon mechanical prime movers. The shift from one fossilized source of energy to another, therefore, is more analogous to shifting plants to forage among hunter-gatherers or species to domesticate among agrarians. These transitions do not fundamentally challenge – but rather support – the continuation of the metabolic logic of the respective socio-ecological regime.

The fact that the introduction of new fossil energy carriers supports the industrial regime is perhaps most clearly illustrated in how the fourth wave of economic development after the Second World War ushered in a massive increase in the metabolic throughput of industrial societies – a historical event now known as "the Great Acceleration" (Steffen et al., 2011, 2015a). Access to and burning of oil in combustion engines facilitated this drastic acceleration of industrial metabolism from 1945 and onward. Oil is an extremely energy-dense source of fossil energy with even higher energy values (40–44 MJ/kg) than coal (18–25 MJ/kg) (Smil, 2016b: 12). In world peripheries and world cores alike, it is hard to overestimate the developmental impact of oil extraction and consumption (Mitchell, 2011). Still today, gross domestic production (GDP) of the world is intimately connected to oil extraction and dissipation – if the world's oil extraction falls 1%, the world's GDP also falls 1% (Ayres, 2016: 382–389). Moreover, the period between 1950 and 2010 saw perhaps the greatest expansion of societal metabolism in world history, with primary energy use increasing 500%, real GDP increasing by 600%, fertilizer consumption reaching 160 million tons per year from practically zero, and world population soaring from 2.5 to 7 billion (Krausmann et al., 2009; Steffen et al., 2015a).

Like any form of fossilized energy source, however, oil is also subjected to diminishing EROI (Hall and Klitgaard, 2012). From the start, oil dependency was questioned based on this fact. The most famous warning came 1956 from Shell-employed geologist Marion King Hubbert who estimated that the US domestic production of petroleum would follow a bell-shaped curve (the Hubbert curve) that he expected would peak in the early 1970s. Hubbert's calculations were first met with skepticism from Shell and the oil industry. However, when the peak in production came in 1971 the tone changed from skepticism to praise. The debate around what today is known as "peak oil" largely sprang from Hubbert's famously accurate predictions of the US oil production peak in the early 1970s. The US domestic peak oil contributed to the 1973 oil crisis by making the US more dependent upon international oil reserves and therefore vulnerable to Arab oil embargos motivated by US interventions in the Middle East. While US demand for oil was eventually saturated, some economists believe that something like the Hubbert curve may be applicable to global oil production today (Hall and Klitgaard, 2012; Ayres, 2016).

Today, however, the total amount of oil produced continues to increase even if so-called "conventional oils" are subjected to diminishing returns. This, some argue, is because oil industries are turning to less lucrative and much dirtier forms of oil, such as tar sands or shale oil, which become cost-effective when the EROI of conventional oils diminish (Ayres, 2016). As the EROI of conventional oils diminish, a larger portion of society's resources must be allocated to oil extraction. Many of the world's governments now provide the necessary resources in the form of subsidies (Erickson et al., 2017). In this way, the extraction and burning of oil continues to increase, despite the global peak in conventional oil production. Crucially, however, this occurs at substantial real-world costs to global and local environments and communities in the form of more frequent devastating weather events, community displacement, and loss of biodiversity (Nduagu and Gates, 2015; Parson and Ray, 2016).

The ecological crisis as a consequence of the Industrial Revolution

With access to energy-dense coal, oil, and fossil gas, human-environmental relations were fundamentally altered. As put by Uddenfeldt (2016: 61, my translation), industrial societies "were not just separated from nature in a philosophical sense, but for real. Free, unbound, and independent." The notion that the development of more advanced technologies made industrial societies free and independent from nature is a surprisingly common assertion. Among ecological modernists, this leads to the contradictory position that "technologies ... have made humans less reliant upon the many ecosystems that once provided their only sustenance, even as those same ecosystems have often been left deeply damaged" (Asafu-Adjaye et al., 2015: 8). But the claim that industrial societies have decoupled from nature is based on a fundamental misunderstanding.

First, industrial societies are utterly dependent upon stocks of fossil energy, raw materials, and fertile lands that either originate from or represent aspects of nature. This nature dependency is in fact increasing. Between 1900 and 2009, industrial capitalism generated a tenfold increase in material extraction (Krausmann et al., 2018). Similarly, between 1970 and 2010 both material and energy throughput increased threefold (UNEP, 2016; BP, 2020). Human appropriation of net primary production (HANPP) is a measure of how much of the world's potential natural vegetation is currently being appropriated by humans. A recent study calculated that 28% of the world's productive surface has been claimed for human land use (agriculture, infrastructure, etc.), a doubling of HANPP in the 20th century (Haberl et al., 2007; Krausmann et al., 2013). The ecological footprint of industrial societies has also drastically increased over the last century, and the sum total of humanity's ecological footprint has been larger than the Earth's biocapacity since the 1970s. The industrial-capital colonization of environments is in turn intimately connected to the massive loss of biodiversity during the 20th century (Barnosky et al., 2011; IPBES, 2019). In particular, the loss of natural habitat due to agricultural intensification has been identified as the biggest threat to global biodiversity (WWF, 2018).

Second, the industrial regime has massive impacts on the natural world through pollution. To quote ecological economist Clive Spash (2017: 9), pollution "is an inevitable part of the economic process, not an avoidable externality that disappears if the prices are 'right'." Any metabolizing system "pollutes" its environment. Typically, however, the pollution of non-human organisms is considered food for fellow species within the ecosystem. In contrast, the millions of artificial chemicals produced in the industrial regime are not the natural foods of other species but are often poisonous. But it is the immense scale and rapidity with which industrial societies dissipate matter-energy that should be the primary cause for concern. The sheer quantities of pollution from the industrial metabolism makes it practically impossible for the ecosystems of the Earth to absorb and sequester the material in time. This is most starkly illustrated by how Earth's nutrient cycles, including the carbon cycle and the nitrogen cycle, have been overburdened by pollution from 20th-century industrial metabolism.

Concerning the nitrogen cycle, the Haber-Bosch process is the single largest cause of the intensification of nitrates into the air and water, something associated with numerous ecological problems including eutrophication and acidification (Smil, 2001). Similarly, the carbon cycle has been affected by the anthropogenic emission of greenhouse gases, leading to concentrations of CO_2 in the atmosphere that now cause a dramatic increase in global temperatures (IPCC, 2014a, 2018). Current levels of CO_2 in the atmosphere must therefore drastically be reduced to avoid catastrophic changes to Earth's life-supporting systems (IPCC, 2018; Steffen et al., 2018). As put by Malm (2018b), "nature comes roaring back" in the form of increased frequency of extreme weather events such as droughts and floods, rising sea levels, loss of coral reefs, and considerably harder conditions for plant cultivation (IPCC, 2018). Already today, the effects of climate change are disproportionally felt by the poor and most vulnerable people in the world system (IPCC, 2014b). To keep global temperatures below 1.5° relative to pre-industrial levels, no more than 275 $GtCO_2$ must be emitted in the period 2016–2100 (Rogelj et al., 2018). Merely during the years 2011–2015, however, as much as 200 $GtCO_2$ were released (ibid). The emissions embodied in the known fossil energy reserves have been estimated to amount to roughly 2,900 $GtCO_2$ (McGlade and Ekins, 2015). With vested economic interests capitalizing on these lucrative fossil assets, it is unlikely that the Earth system will be spared further detrimental changes unless a radical transition toward another socio-ecological regime occurs (Foster, 2013).

The impacts of all past civilizations fade in comparison to the industrial regime. No other energy transformation in history has had such an all-pervasive effect on the natural world as the Industrial Revolution. It is not difficult to understand the industrial regime's problematic relation to the environment against the definition of industrialism as an inclination toward the mass production of commodities and capitalism as a global system of exchange driven by the social aspiration to accumulate and profit *ad infinitum*. As we have seen, nothing comes from nothing

(Chapter 2). This means that the ever-increasing mass production of commodities is coupled with an ever-increasing mass extraction of the necessary matter-energy (Adams, 1982). At the other end of the social metabolism, mass pollution also has a considerable effect upon the biosphere in the form of accumulation of numerous artificial chemicals and greenhouse gases. In this sense, the development experiences by people in the world core over the 19th and 20th centuries is inseparable from the destruction and alteration of natural environments that are now threatening the continuation of life on an unprecedented scale.

In this context it is absurd to suggest that industrial capitalism is becoming increasingly independent from nature. In the midst of a historical situation characterized by increasing extraction, production, consumption, and pollution, i.e., metabolizing, of an ever-increasing amount of matter-energy, the industrial metabolism is now fundamentally altering nature's nutrient cycles that have been stable for thousands of years. It therefore seems more correct to say that societies' dependence upon nature is *rapidly increasing*. Nature's processes are now so deeply affected by industrial capitalism that it has raised debates regarding humans as the major force in a new geological epoch (cf. Malm and Hornborg, 2014). Even if ecomodernists think of the Anthropocene as a historically "great" achievement (Asafu-Adjaye et al., 2015), the unequivocal hallmark of this epoch is the destruction of millions of years' worth of evolution and biodiversity through human interference in Earth's life-supporting systems. While the tendency to alter environments at the point of civilizational collapse has occurred in the past, never before has the Earth system been altered so radically by any human metabolic interaction. This, ultimately, is a situation uniquely attributable to the industrial regime.

World division of labor as a consequence of the Industrial Revolution

The industrial regime has facilitated social and economic improvements that have generated greater freedom and independence in certain aspects of life compared to some earlier societies. Never before have so many been able to consume so many goods and enjoy so much material wealth or been so educated. Never before was it possible to travel vast distances at incredible speeds or communicate with another person on the other side of the globe in real time. In part due to modern medicine, human health conditions have now also returned to levels equivalent to those of the Upper Pleistocene (Gowdy, 2020). Politically, industrialization and access to fossil fuels have been shown to correlate with social revolutions, some of which swept away despotism and introduced more democratic forms of governance (Mitchell, 2011). The world, some insist, is getting better and better (e.g., Pinker, 2018; Rosling, 2018). But this is not the case for everyone. With all the evidence at hand, industrial capitalism is still characterized by a world division of labor that shares burdens and benefits unequally (Harrington et al., 2016; Oxfam, 2017; Chakraborty et al., 2018). Crucially, this discrepancy is not due to differences in technological progress or cultural sophistication, but a result of the historically developed condition with roots in Western colonial expansion and capitalist relations of production (Wolf, 2010[1982]; Bunker, 1985; Mitchell, 2011).

Despite all its advantages to beneficiary regions and classes, industrial capitalism does not challenge the world division of labor. Why not? The short answer is that this world division is a *sine qua non* of the regime itself. In other words, without this division, it would grind to a halt. Global discrepancies in environmental burdens and benefits are not simply a consequence of the industrial regime, but a necessary condition for its continuation. Without a means to access highly dense forms of energy and massive quantities of high-quality raw materials scattered throughout the world, it is unlikely the world-economic cores would be able to maintain a continually increasing level of matter-energy throughput. Industrial capitalism, in the end, does not challenge the world division of labor because it would then undermine its own capacity to develop technologies that could penetrate nature and social life in novel ways to establish new market outlets for capital investment and profit. This means that ecologically unequal exchange is not a side effect of industrial metabolism, but a necessary aspect of its reproduction (Hornborg, 2013; cf. Andersen, 2013).

The mechanisms by which it is possible to maintain this relation include (a) securing necessary resources through military means or (b) trading resources under price differences that provide favorable terms of trade in a biophysical sense (see Chapter 4; Hornborg, 1998; Pérez-Rincón, 2006). Examples of the former include the US invasion of Iraq in 2003 that secured US access to lucrative oil reserves. Timothy Mitchell (2011) has brilliantly shown how democratic governance in core regions of the world are dependent upon anti-democratic governance in peripheral regions, such as Iraq, that facilitates the flow of petroleum—and so its social-political benefits –from periphery to core. Examples of the latter include a wide range of historical and contemporary cases of ecologically unequal exchange as well as neoliberal schemes such as "structural adjustment programs" that deregulate and force open publicly managed environments and resources of impoverished and indebted countries to the world market at low prices (e.g., Jorgenson, 2010; Frey et al., 2019).

While this social condition generates enormous wealth in core regions of the world, it simultaneously impedes wealth generation in peripheral regions (Jorgenson et al., 2009). We can therefore conclude that fossil fuels both extend and deepen capitalist relations of production through a process that is perhaps best described as "ecological imperialism." Ultimately, then, the industrial regime reproduces the same world division of labor upon which it was founded.

The modern worldview as a consequence of the Industrial Revolution

For some time, anthropologists have insisted that people in world-economic cores are just as susceptible to magical beliefs as non-modern peoples (Stivers, 1999; Hornborg, 2016). This insight challenges the conventional self-image of modern culture as having been liberated from supposedly false beliefs concerning the world (Hornborg, 2015: 866). The industrial regime is no historical exception when it comes to questionable cultural imaginaries, even if the dominant narrative of the Industrial Revolution relies upon this claim (Landes, 1969). Studies in emerging fields such as "energy humanities" have shown how irrational cultural imaginaries are formed and upheld through processes reliant upon fossil fuels in industrial societies (Boyer, 2011; Love and Isenhour, 2016; Wilson et al., 2017). Questions regarding how fossil fuels have affected modern culture include a broad range of issues, from the modern notion of freedom (Huber, 2015) to something as innocent as the celebration of New Year (Uddenfeldt, 2016). As suggested by Wilson et al. (2017: 8), "just as politics has been shaped by and in reaction to oil, so, too, have many of our most important concepts and theories." For the purpose of this book, I will highlight only two such categories, the immaterial conception of the economy and the notion of technologies as inherently productive.

Ecological economists have for a long time critiqued neoclassical economists for failing to recognize biophysical nature as an essential foundation for the economic process (Georgescu-Roegen, 1971, 1975). Scholars from other fields have more recently also pointed out that the neoclassical school of economics emerged in tandem with the increased access to and burning of fossil fuels during the late 19th and early 20th centuries (Mitchell, 2011; Wilson et al., 2017). With the access to extremely lucrative and seemingly nature-independent sources of fossil fuels, economic theories became less concerned with physical production factors (e.g., land, labor, energy, materials) (Mitchell, 2011). This represents a paradox because the rise of the neoclassical school around 1860–1870 (the Marginalist Revolution) is commonly understood as based upon insights from the physical sciences. The crucial issue, as pointed out by ecological economists and historians, is that the insights from thermodynamics were only adopted *metaphorically*, with little interest for understanding the biophysical basis of the economy (Mirowski, 1989). To Jevons, for example, the economic notion of utility was understood as parallel to the physicist's notion of gravity, and value as parallel to energy (Mirowski, 1989: 219). The failure to realize how the economic process is *actually* subjected to the law of entropy, Georgescu-Roegen (1975) argued, was intimately connected to the mechanistic-epistemological inclination in neoclassical economics, through which the economy is understood as a perpetual motion machine operating independently from an outside environment. Thus, fossil fuels, the Earth's species, and nature's biogeochemical cycles were misconstrued as somehow "external" to the economy, all while economists took pride in the scientific basis of their discipline.

The notion of the economy as nature-independent arose in tandem with the notion that mechanical artifacts are inherently productive. If the wealth generated by the economy does not originate from labor, land, or material resources, it must have some other origin. As we have seen, instrumental descriptions attribute the productive potential to human scientific knowledge (e.g., Landes, 1969). Even today, it is common to attribute the productive potential of machines to scientific knowledge or even to the machines themselves rather than attributing it to the natural origin of the matter-energy that it transforms. The failure to realize the distinction between "energy converters" and "energy carriers/sources" still confuses modern economic analyses of the Industrial Revolution and technological progress (Giampietro and Mayumi, 2008; Hornborg and Roos, 2022). This is a situation in which the productive potential of technologies appears to originate only from the prime mover assembled in society, when it is in fact also contingent upon the primary source of energy derived from nature. One of the results of this, as we have seen, is the common belief that "innovation" or "engineering design," quite independent of any environmental concern, is a sufficient condition for developing new technologies and generate more wealth.

The immaterial conception of the economy and the notion of technologies as inherently productive are both symptoms of alienation, which makes the impossibility of infinite economic growth seem possible. As suggested by Vadén and Salminen (2018: 41), "one characteristic of modernity is the blindness to its material conditions." Separated from directly engaging in the reproduction of their communities, modern urban peoples do not require in their daily life a notion of how the commodities they rely upon are produced, or, even that they are produced. From the alienated gaze of urban moderns, wealth does not originate from nature, but from the sources, which they can themselves identify as relevant in their lives: "innovation," "design," and, above all, "money." This is in stark contrast to the historical peasant's "image of the limited good" in which all wealth originated from a finite nature (Foster, 1965). At least since the 1960s, researchers and environmental activists have pointed out the pitfalls of not considering the limits, thresholds and boundaries of the Earth and our immediate environments. But the substance of these historically informed wisdoms evaporate when the immaterial conception of the economy and the notion of inherently productive technology inform the solutions: More economic growth and more advanced technologies. One important result of this is the rise of the notion that it is possible to leave the industrial regime without setting off wider social-ecological changes that challenge its technological development.

Solar power in the new metabolic regime

As stated in the introduction to this chapter, some scholars point out that leaving the fossil regime implies a metabolic transformation that is comparable to the Neolithic and Industrial Revolutions. We now have a historically rooted understanding of what this actually means, beyond merely implying a shift in the dominant energy carrier. I have given an overview of how the Industrial Revolution gave rise to historically novel social, cultural, and ecological relations. What social, cultural, and ecological relations and conditions are we then to expect from the now much-anticipated transformation away from fossil energy? To be sure, given the major changes associated with such major energy transformations, "it is probably as difficult for us to imagine a sustainable society as it was for people in the 16th century to imagine the industrial society today" (Haberl et al., 2011: 11). We have no way of knowing exactly what the future will look like. Most likely, history will make fools of anyone who attempts any such prediction. There are, nevertheless, current processes and events that can be put under scrutiny to begin to understand some of the prerequisites of the new social metabolism.

Putting aside the dip in fossil energy throughput during the first year of the corona pandemic, there is little to suggest that the industrial regime based on fossil fuels is undergoing a process of fundamental metabolic transformation (IEA, 2021). Even if the consumption of electricity generated through renewable energy technologies is increasing, fossil energy carriers still supply 80%–90% of the world's primary energy (Smil, 2016c). Since diminishing energy throughput is antithetical to corporate investments in fossil energy reserves and social-political complexity at large, it also seems reasonable to expect that the push for higher levels of fossil energy throughput will persist if no radical action is taken. This becomes clear in the light of recent studies, which show how the installation of renewable energy technologies tend to add to, not replace, already fossil-dominated energy mixes (see Chapter 1). We might nevertheless be witnessing the beginning of an energy addition in which the economic imperative of endless accumulation pushes society to increase both fossil energy and renewable energy simultaneously.

There are several concerns that call into question whether renewable energy technologies can support industrial levels of energy-matter throughput in the absence of fossil energy subsidies or a world division of labor. The central question today concerning energy transformation is therefore whether the development of renewable energy technologies is currently in the process of continuing, transcending, or reversing the industrial regime. Let us look at each in turn.

Continuing the industrial regime: The conventional position is that the increased installation of renewable energy technologies represents the emergence of a more sustainable human-environmental relation that will continue the industrial regime. In this view, declining prices of renewable energy and increasing technological efficiencies are referred to as signs of the emergence of a sustainable and universal form of industrial capitalism (see, e.g., UNDP, 2016). However, in relation to the technological continuum, the social-ecological prerequisites for the existence of renewable energy technologies are systematically ignored or downplayed. In effect, the installation of renewable energy technologies commonly take on the discursive role of a "technological fix" that bypasses the necessity for significant social, cultural, or biophysical changes that are otherwise associated with transforming the industrial regime. Arguably, this position only holds as far as colonialism and the metabolism of fossil energy are analytically separated from industrialism. In conventional interpretations, the Industrial Revolution was made possible historically through British engineering design applied to harness natural forces (see, e.g., Landes, 1969). By extension, scientific knowledge is considered the causal prime mover of energy transformations. Since industrialism itself arose out of human ingenuity, it follows that it can be made sustainable through the application of better scientific knowledge and technological design. Because of this, no material prerequisite, such as a particular energy source or a particular social relation is seen as necessary for the continuation of industrialism. Designs and technological improvements of renewable energy technologies, nuclear reactors, and/or fossil energy converters are considered the novel scientific knowledge upon which the new metabolism will be based. In this sense, current levels of production and consumption can be extended ad infinitum without further harm to the biosphere or people throughout the world.

Transcending the industrial regime: The second position sees the conditions and trajectories of the current historical moment as a moral obligation to transform industrial capitalism into a new metabolism that maintains industrial levels of matter-energy throughput under more just social relations. This stance argues that the post-fossil era must necessarily be post-capitalistic. This is shown, in part, by demonstrating how fossil energy carriers were necessary for the emergence and lock-in of the capitalist division of labor within Britain during the Industrial Revolution (Huber, 2008; Malm, 2016a, b). As argued by Andreas Malm (2016a, b), early British cotton manufacturers employed fossil fuel-propelled machinery because it provided a means for superior control over workers through its (seemingly) landscape-independent character.[8] Still today, Malm (2013, 2016a, b, 2018a, b) argues, capitalism cannot exist without fossil energy because its biophysical characteristics are essential for the mobility of transnational corporations relying upon the relocation of factories to regions with low prices for manufacturing in order to stay competitive on the world market. It follows that if we wish to leave fossil energy in the ground to mitigate environmental problems such as climate change, capitalism needs to be radically transformed.

This transformation can occur through the development and installation of renewable energy technologies that are qualitatively different from fossil energy carriers by virtue of being more integrated in landscapes (Malm, 2016a, b, 2018a, b). Renewable energy technologies, as such, are antithetical to the capital accumulation of transnational corporations because these corporations necessitate fossil fuels to move extraction processes and manufacturing facilities across the world with impunity. It cannot be expected that actors, whose primary interest consists in accumulating capital, should develop renewable energy technologies, since maintaining high rates of profit requires low wages in the world economy, which can only be facilitated by fossil energy carriers. What is needed instead is the active and purposeful intervention of powerful states that can connect the comparative advantage of different national energy-landscapes in international super grids harnessing renewable or nuclear sources of energy. Through central planning, a more sustainable and greener metabolism can arise that challenges the social-ecological relations of industrial capitalism (Schwartzman, 1996). Crucial to this position is that the energy transformation will only occur through an increased matter-energy throughput (Schwartzman, 2012; Phillips, 2015). This change will necessarily arise from a social telos connected to a revolutionary class that considers the social metabolism not as an engine for profit or capital accumulation (e.g., Huber, 2019). The industrial regime will therefore be transcended simultaneously as the social relations of capitalism. In this sense, through a purposeful and rapid revolution, industrial levels of production and consumption can be extended justly throughout the world without further harm to the biosphere.

Reversing the industrial regime: The third position considers the industrial regime as a historical parenthesis that is fundamentally contingent upon fossil energies that are now showing signs of diminishing energy returns. This is supported by a number of points showing how leaving fossil fuels in the ground will reintroduce some of the major characteristics of the agrarian regime.

Similar to the second position, those who think that the industrial regime is being reversed recognizes the landscape-dependent character of renewable energy sources as a feature that sets them apart from fossil energies. As we have seen, the land – and energy efficiencies of renewable energy sources such as solar PVs and biofuels are low in comparison to fossil energy carriers (see Chapters 1 and 5). This also applies to unconventional oils, such as tar sands and shale oil (Cleveland and O'Connor, 2011; Smil, 2015). Compared to the industrial regime, the new metabolism is therefore predicted to be vastly more dependent on direct surface areas for maintaining current (or higher) levels of energy throughput (Scheidel and Sorman, 2012). The same applies to both mineral and energy requirements.

Contrary to the second position, the comparatively large biophysical demand to generate renewable energy is not understood as antithetical to global capitalism. The point here is a subtle one: While fossil fuels reduced the costs of relocating factories to regions with low wages, it is the resultant wage and price differences in the economy, not fossil fuels per se, which determine the continued success of capital accumulation. As history shows, there are other means for creating and upholding regional or global price differences, including enslavement or indebtment. To the extent that current price differences are maintained (or deepened), the real costs for generating renewable energy is a relative issue contingent upon wages and prices in a particular region. Since the industrial regime arose from colonialism and the price differences enforced through slavery and military domination, there is a possibility that capitalist relations of production will continue in a new metabolism energized again by renewable energy sources.

A throughput as large as in today's affluent societies is likely to require enormous amounts of resources. Without access to subterranean stocks of energy (fossil fuels), these resources must be extracted from current ecosystems located above the Earth's crust. This leads to the question whether current levels of consumption will be maintained in core regions at the expense of peripheral regions of the world. This would effectively reverse the Industrial Revolution by making core countries increasingly dependent upon embodiments of labor and resources in international trade. The larger relative land requirements per unit of energy implied by biofuels and solar PV energy indicate that the new metabolism may necessitate ecologically unequal exchange (see, e.g., Hermele, 2014; Hornborg, 2014; Hornborg et al., 2019). To the extent that this is true, it is unlikely that a matter-energy throughput as large as in today's affluent societies is an ambition that is environmentally sustainable or socially just, even if it propelled by renewable energy. This non-universality contrasts to the first and second positions' visions of a new metabolism that will distribute high energy-matter throughput equally throughout the world through the development of renewable industrial mega-projects.

At times it is unclear whether the different positions above argue what their respective authors *want* to happen or what they sincerely *think* will happen. This may be thought of as the distinction between a utopian and an analytical lens. Both lenses must arguably be combined if we wish to establish an at once desirable and credible vision of the new metabolism. That is to say, any social-technical aspiration must be understood together with the social-ecological conditions necessary to realize it. In the end, "people make their own history, but not exactly as they please" (Foster et al., 2010: 38). For the remainder of this book, I will show how the emerging metabolic regime – to the extent that it relies on solar PV technology – generates conditions that are best described as reversing the Industrial Revolution. In the following chapters, then, the contrast between the socially articulated visions of solar technology and the social-ecological reality of solar technology will become more prominent to the reader. Without revealing too much of the coming chapters, it will suffice to say that people are now making what we may call "solar history," but not exactly as they envisioned.

Notes

1 As we saw in Chapter 2, humans cannot choose to end this transformative interaction if they wish to survive in nature, since the reproduction of society necessitates a matter-energy exchange, a Stoffwechsel, which always implies a transformation of nature.

2 What in historical materialism are called the "ancient mode of production" and the "feudal mode of production" can be understood both to be encompassed within the "agrarian socio-ecological regime" because they are both metabolically based on energy captured through cultivation of plants. Modes of production have at times changed simultaneously as transitions in social-ecological regimes. Hypothetically, therefore, new social-ecological regimes might imply new modes of production in the Marxist sense, but not necessarily the other way around (see Fischer-Kowalski et al., 2019). As such, it seems likely that changes in social relations alone do not determine changes in socio-ecological regimes. However, the opposite might hold some historical truth. As argued by York and Mancus (2009), this difference in periodization between "critical human ecology" and classical historical materialism is rooted in the former's acceptance of an ahistorical understanding of nature, such as exemplified by natural laws.

3 I distinguish "social metabolism" from "social-ecological regime." Whereas "social metabolism" denotes a socially organized exchange of matter-energy with the environment, a "social-ecological regime" denotes the social metabolism and the cultural imaginaries to support it.

4 Large-scale industrial operations to manufacture machines, construct railways, extract coal, and transport commodities across the world required a substantial amount of workers and raw materials. Both the food for this workforce (including both wage laborers and slaves) and many of the raw materials themselves implied substantial amounts of biomass. This includes biomass in the form of sugar from South and Central America, cotton from North America, and wheat and wood from the Baltic (see below). As concluded by Inikori (2002: 478), "the claim, that technological development... caused the growth of overseas sales instead of the other way round, is contrary to the clear evidence from the northern countries that led the technological change in cottons, woolens, and metals." In short, technological development and industrialization was contingent on biomass and raw material importation.

5   These calculations only account for the land occupied by the mine itself (10,000 m^2), i.e., not the land embodied in the mechanical infrastructure, the capital, or the labor.
6   The transition to fossil fuels was not simply a transition from biomass to coal, but also a transition from water energy to fossil energy in early key industries such as the cotton manufacturing industry (Malm, 2016a, b). It is nevertheless crucial that we understand the extraction of fossil fuels and industrial technology as an alternative means for increasing matter-energy throughput that is distinct from the territorial expansion in the agrarian regime. This, however, does not imply that the conquest of territories was halted with the Industrial Revolution. Quite the contrary, the increased access to energy was still contingent upon an increased access to materials extracted from environments in remote lands for building and maintaining highly complex forms of (energy) infrastructures in the core (Bunker and Ciccantell, 2005a; Hornborg, 2006). In this sense, fossil energy was not substitutive of, but additive to, colonial strategies for matter-energy appropriation (Barca, 2011; Hornborg, 2015). As is becoming increasingly clear in scholarly work, fossil energies did not prevent the logic of capitalism, rather it cemented capitalist relations of production, wage labor, and the spread of the world market.
7   Defined as components converting energy sources into motive power, also known as "energy converters" (see Giampietro and Mayumi, 2008).
8   Fossil fuels, once they are extracted, are highly transportable energy carriers that can be used to power machinery in almost any location. This sets them apart from biomass, waterpower, and wind power. This landscape-independent character, however, is illusory, since the effects of fossil fuels upon the landscape is today returning in the form of extreme weather events (not to mention the desolate landscapes generated from the extraction of coal or tar sands).

References

Abu-Lughod, Janet L. 1989. *Before European hegemony: The world system A.D. 1250–1350.* New York, NY and Oxford: Oxford University Press. Adams, Richard N. 1982. *Paradoxical harvest: Energy and explanation in British history, 1870–1914.* Cambridge: Cambridge University Press.

Andersen, Otto. 2013. *Unintended consequences of renewable energy: Problems to be solved.* London, England: Springer.

Asafu-Adjaye, John, Linus Blomqvist, Stewart Brand, Barry Brook, Ruth DeFries, Erle Ellis, Christopher Foreman, et al. 2015. *An ecomodernist manifesto.*www.ecomodernism. org.

Ayres, Robert. 2016. *Energy, complexity, and wealth maximization.* Cham, Germany: Springer.

Barca, Stefania. 2011. Energy, property, and the industrial revolution narrative. *Ecological Economics* 70: 1309–1315. https://doi.org/10.1016/j.ecolecon.2010.03.012.

Barnosky, Anthony D., Nicholas Matzke, Susumu Tomiya, Guinevere O. U. Wogan, Brian Swartz, Tiago B. Quental, Charles Marshall, et al. 2011. Has the earth's sixth mass extinction already arrived? *Nature* 471: 51–57. https://doi.org/10.1038/nature09678.

Boyer, Dominic. 2011. Energopolitics and the anthropology of energy. *Anthropology News* 52(5): 5–7.

BP. 2020. *BP statistical review of world energy.* 69th ed. London: BP.

Bridge, Gavin. 2010. The hole world: Scales and spaces of energy extraction. *New Geographies 2: Landscapes of Energy* 2: 43–51.

Bunker, Stephen G. 1985. *Underdeveloping the Amazon: Extraction, unequal exchange, and the failure of the modern state*. Chicago and London, England: University of Chicago Press.

Bunker, Stephen G. 2007. Natural values and the physical inevitability of uneven development under capitalism. In *Rethinking environmental history: World-system history and global environmental change*, edited by Alf Hornborg, J. R. McNeill, and Joan Martinez-Alier, 239–358. Lanham, MD and New York, NY: AltaMira.

Bunker, Stephen G., and Paul S. Ciccantell. 2005. *Globalization and the race for resources*. Baltimore, MD: Johns Hopkins University Press.

Chakraborty, Jayjit, Timothy W. Collins, and Sara E. Grineski. 2018. Exploring the environmental justice implications of hurricane Harvey flooding in greater Houston, Texas. *American Journal of Public Health* 109(2): 244–250. https://doi.org/10.2105/AJPH.2018.304846.

Clark, Brett, and John B. Foster. 2009. Ecological imperialism and the global metabolic rift: Unequal exchange and the guano/nitrates trade. *International Journal of Comparative Sociology* 50(3–4): 311–334. https://doi.org/10.1177/0020715209105144.

Cleveland, Cutler J., and Peter A. O'Connor. 2011. Energy return on investment (EROI) of oil shale. *Sustainability* 3: 2307–2322. https://doi.org/10.3390/su3112307.

Crosby, Alfred W. 1986. *Ecological imperialism: The biological expansion of Europe, 900–1900*. Cambridge: Cambridge University Press.

Debeir, Jean-Claude, Jean-Paul Deléage, and Daniel Hémery. 1991. *In the servitude of power: Energy and civilization through the ages*. London, England and New Jersey: Zed Books.

Dukes, Jeffrey S. 2003. Burning buried sunshine: Human consumption of ancient solar energy. *Climatic Change* 61(1): 31–44. https://doi.org/10.1023/A:1026391317686.

Erickson, Peter, Adrian Down, Michael Lazarus, and Doug Koplow. 2017. Effects of subsidies to fossil fuel companies on United States crude oil production. *Nature Energy* 2: 891–898. https://doi.org/10.1038/s41560-017-0009-8.

Fischer-Kowalski, and Helmut Haberl. 2007. Conceptualizing, observing and comparing socioecological transitions. In *Socioecological transitions and global change*, edited by Marina Fischer-Kowalski, and Helmut Haberl, 1–30. Cheltenham, England and Northampton, MA: Edward Elgar.

Fischer-Kowalski, Marina, Elena Rovenskaya, Fridolin Krausmann, Irene Pallua, and John R. Mc Neill. 2019. Energy transitions and social revolutions. *Technological Forecasting and Social Change* 138: 69–77. https://doi.org/10.1016/j.techfore.2018.08.010.

Foster, George M. 1965. Peasant society and the image of limited good. *American Anthropologist* 67: 293-314.

Foster, John Bellamy. 2000. *Marx's ecology: Materialism and nature*. New York: Monthly Review Press.

Foster, John Bellamy. 2013. The fossil fuels war. *Monthy Review*. 01-09-2013. https://monthlyreview.org/2013/09/01/fossil-fuels-war/. Accessed 30-07-2020.

Foster, John Bellamy, Brett Clark, and Richard York. 2010. *The ecological rift: capitalism's war on the earth*. New York, NY: Monthly Review Press.

Frank, Andre G. 1966. The development of underdevelopment. *Monthly Review* 18(4): 17–31.

Frey, Scott R., Paul K. Gellert, and Harry F. Dahms, eds. 2019. *Ecologically unequal exchange: Environmental justice in comparative and historical perspective*. Cham, Germany: Palgrave Macmillan.

Geels, Frank W. 2012. A socio-technical analysis of low-carbon transitions: Introducing the multi-level perspective into transport studies. *Journal of Transport Geography* 24: 471–482. https://doi.org/10.1016/j.jtrangeo.2012.01.021.

Geels, Frank W. 2014. Regime resistance against low-carbon transitions: Introducing politics and power into the multi-level perspective. *Theory, Culture and Society* 31(5): 21–40. https://doi.org/10.1177/0263276414531627.

Gellert, Paul K., and Paul S. Ciccantell. 2020. Coal's persistence in the capitalist world-economy: Against teleology in energy 'transition' narratives. *Sociology of Development* 6(2): 194–221. https://doi.org/10.1525/sod.2020.6.2.194.

Georgescu-Roegen, Nicholas. 1971. *The entropy law and the economic process*. London, England: Oxford University Press.

Georgescu-Roegen, Nicholas. 1975. Energy and economic myths. *Southern Economic Journal* 41(3): 347–381. https://doi.org/10.2307/1056148.

Giampietro, Mario, and Kozo Mayumi. 2008. Complex systems thinking and renewable energy systems. In *Biofuels, solar and wind as renewable energy systems*, edited by David Pimentel, 173–213. Dordrecht, the Netherlands: Springer.

Gowdy, John M. 2020. Our hunter-gatherer future: Climate change, agriculture and uncivilization. *Futures* 115: 102488. https://doi.org/10.1016/j.futures.2019.102488.

Haberl, Helmut, K. Heinz Erb, Fridolin Krausmann, Veronica Gaube, Alberte Bondeau, Christoph Plutzar, Simone Gingrich, Wolfgang Lucht, and Marina Fischer-Kowalski. 2007. Quantifying and mapping the human appropriation of net primary production in earth's terrestrial ecosystems. *PNAS* 104(31): 12942–12947. https://doi.org/10.1073/pnas.0704243104.

Haberl, Helmut, Marina Fischer-Kowalski, Fridolin Krusmann, Joan Martinez-Alier, and Verena Winiwarter. 2011. A socio-metabolic transition towards sustainability? Challenges for another great transformation. *Sustainable Development* 19(1): 1–14. https://doi.org/10.1002/sd.410.

Hagens, Nathan J. 2020. Economics for the future—Beyond the superorganism. *Ecological Economics* 169: 106520. https://doi.org/10.1016/j.ecolecon.2019.106520.

Hall, Charles A. S., and Kent A. Klitgaard. 2012. *Energy and the wealth of nations: Understanding the biophysical economy*. New York, NY: Springer.

Hall, Charles A. S., Jessica G. Lambert, and Stephen B. Balogh. 2014. EROI of different fuels and the implications for society. *Energy Policy* 64: 141–152. https://doi.org/10.1016/j.enpol.2013.05.049.

Harrington, Luke J., David J. Frame, Erich M. Fischer, Ed Hawkins, Manoj Joshi, and Chris D. Jones. 2016. Poorest countries experience earlier anthropogenic emergence of daily temperature extremes. *Environmental Research Letters* 11: 055007. https://doi.org/10.1088/1748-9326/11/5/055007.

Harris, Marvin. 1997. *Culture, people, nature*. New York, NY: Longman.

Harvey, David. 2003. *The new imperialism*. Oxford and New York, NY: Oxford University Press.

Headrick, Daniel R. 2009. *Technology: A world history*. Oxford and New York, NY: Oxford University Press.

Hermele, Kenneth. 2014. *The appropriation of ecological space: Agrofuels, unequal exchange and environmental load displacement*. Abingdon, Oxon and New York, NY: Routledge.

Hornborg, Alf, and Andreas Roos. 2022. Technology as capital: Challenging the illusion of the neutral machine. In *Capitalising on the sun: Critical perspectives on the global solar economy*, edited by Dustin Mulvaney, Jamie Cross, and Benjamin Brown. Baltimore, MD: Johns Hopkins University Press. Forthcoming.

Hornborg, Alf, Gustav Cederlöf, and Andreas Roos. 2019. Has Cuba exposed the myth of "free" solar energy? Energy, space, and justice. *Environment and Planning E: Nature and Space* 2(4): 989–1008. https://doi.org/10.1177/2514848619863607.

Hornborg, Alf. 1998. Towards an ecological theory of unequal exchange: Articulating world systems theory and ecological economics. *Ecological Economics* 25(1): 127–136. https://doi.org/10.1016/S0921-8009(97)00100-6.

Hornborg, Alf. 2006. Footprints in the cotton fields: The Industrial Revolution as time-space appropriation and environmental load displacement. *Ecological Economics* 59(1): 74–81. https://doi.org/10.1016/j.ecolecon.2005.10.009.

Hornborg, Alf. 2010. *Myten om maskinen: Essäer om makt, modernitet och miljö.* Borgå, Finland: Daidalos.

Hornborg, Alf. 2013. *Global ecology and unequal exchange: Fetishism in a zero-sum world.* London, England: Routledge.

Hornborg, Alf. 2014. Why solar panels don't grow on trees: Technological utopianism and the uneasy relation between Marxism and Ecological Economics. In *Green utopianism: Perspectives, politics and micro-practices*, edited by Johan Hedrén, and Karin Bradley, 76–97. New York, NY: Routledge.

Hornborg, Alf. 2015. Industrial societies. *International Encyclopedia of the Social and Behavioral Sciences* 11: 863–867. https://doi.org/10.1016/B978-0-08-097086-8.12200-7.

Hornborg, Alf. 2016. *Global magic: Technologies of appropriation from ancient Rome to Wall Street.* London, England: Palgrave Macmillan.

Huber, Matthew T. 2008. Energizing historical materialism: Fossil fuels, space and the capitalist mode of production. *Geoforum* 40(1): 105–115. https://doi.org/10.1016/j.geoforum.2008.08.004.

Huber, Matthew. 2015. Theorizing energy geographies. *Geography compass* 9(6): 327-338. https://doi.org/10.1111/gec3.12214.

Huber, Matthew T. 2019. Ecological politics for the working class. *Catalyst* 3(1).

Huber, Matthew T., and James McCarthy. 2017. Beyond the subterranean energy regime? Fuel, land use and the production of space. *Transactions of the Institute of British Geographers* 42(2): 655–668. https://doi.org/10.1111/tran.12182.

IEA. 2020. *World energy outlook 2020.* Paris, France: IEA Publications.

IEA. 2021. Global energy review: CO_2 emissions in 2020. Understanding the impacts of covid-19 on global CO_2 emissions. *IEA*. Published 02-03-2021. https://www.iea.org/articles/global-energy-review-co2-emissions-in-2020. Accessed 12-03-2021.

Inikori, Joseph E. 2002. *Africans and the industrial revolution in England: A study in international trade and economic development.* Cambridge: Cambridge University Press.

IPBES. 2019. *Summary for policymakers of the global assessment report on biodiversity and ecosystem services of the intergovernmental science-policy platform on biodiversity and ecosystem services.* Bonn, Germany: IPBES Secretariat.

IPCC. 2014a. *Climate Change 2014: Synthesis report.* Contribution of working groups I, II and III to the fifth assessment report of the intergovernmental panel on climate change. IPCC, Geneva, Switzerland.

IPCC. 2014b. Livelihoods and poverty. In *Climate change 2014—Impacts, adaptation and vulnerability: Part A: Global and sectoral aspects: Working group II contribution to the IPCC fifth assessment report*, 793–832. Edited by Olsson, L., M. Opondo, P. Tschakert, A. Agrawal, S.H. Eriksen, S.Ma, L.N. Perch, and S.A. Zakieldeen. Cambridge: Cambridge University Press

IPCC. 2018. Summary for policymakers. In *Global warming of 1.5°C. An IPCC special report on the impacts of global warming of 1.5°C above pre-industrial levels and related global greenhouse gas emission pathways, in the context of strengthening the global response to the threat of climate change, sustainable development and efforts to eradicate poverty.* Geneva, Switzerland: IPCC, edited by Masson-Delmotte, V., P Zhai, H.O. Pörtner, D. Roberts, J.

Skea, P.R. Shukla, A. Pirani, et al. 3-24. Cambridge and New York: Cambridge University Press. https://doi.org/10.1017/9781009157940.00.

Jevons, William Stanley. 1865. *The coal question. An inquiry concerning the progress of the nation and the probable exhaustion of our coal-mines.* London and Cambridge, England: Macmillan. https://archive.org/details/coalquestionani00jevogoog.

Jorgenson, Andrew K. 2010. World-economic integration, supply depots, and environmental degradation: A study of ecologically unequal exchange, foreign investment dependence, and deforestation in less developed countries. *Critical Sociology* 36(3): 453–477. https://doi.org/10.1177/0896920510365204.

Jorgenson, Andrew K., Kelly Austin, and Christopher Dick. 2009. Ecologically Unequal Exchange and the Resource Consumption/Environmental Degradation Paradox: A Panel Study of Less-Developed Countries, 1970-2000. *International journal of comparative sociology* 50(3-4): 236-284. https://doi.org/10.1177/0020715209105142.

Kallis, Giorgios, and Richard B. Norgaard. 2010. Coevolutionary ecological economics. *Ecological Economics* 69: 690–699. https://doi.org/10.1016/j.ecolecon.2009.09.017.

Krausmann, Fridolin, Christian Lauk, Willi Haas, and Dominik Wiedenhofer. 2018. From resource extraction to outflows of wastes and emissions: The socioeconomic metabolism of the global economy, 1900–2015. *Global Environmental Change* 52: 131–140. https://doi.org/10.1016/j.gloenvcha.2018.07.003.

Krausmann, Fridolin, Helga Weisz, and Nina Eisenmenger. 2016. Transitions in sociometabolic regimes throughout human history. In *Social ecology: Society-nature relations across time and space*, edited by Helmut Haberl, Marina Fischer-Kowalski, Fridolin Krausmann, and Verena Winiwarter, 63–92. New York, NY: Springer.

Krausmann, Fridolin, Karl-Heinz Erb, Simone Gingrich, Helmut Haberl, Alberte Bondeau, Veronika Gaube, Christian Lauk, Christoph Plutzar, and Timothy D. Searchinger. 2013. Global human appropriation of net primary production doubled in the 20th century. *PNAS* 110(25): 10324–10329. https://doi.org/10.1073/pnas.1211349110.

Krausmann, Fridolin, Simone Gingrich, Nina Eisenmenger, Karl-Heinz Erb, Helmut Haberl, and Marina Fischer-Kowalski. 2009. Growth in global materials use, GDP and population during the 20th century. *Ecological Economics* 68(10): 2696–2705.

Landes, David S. 1969. *The unbound Prometheus: Technological change and industrial development in Western Europe from 1750 to the present.* Cambridge and New York, NY: Cambridge University Press.

Lenton, M. Timothy, Peter-Paul Pichler, and Helga Weisz. 2016. Revolutions in energy input and material cycling in earth history and human history. *Earth System Dynamics* 7: 353–370. https://doi.org/10.5194/esd-7-353-2016.

Love, Thomas, and Cindy Isenhour. 2016. Energy and economy: Recognizing high-energy modernity as a historical period. *Economic Anthropology* 3: 3–16. https://doi.org/10.1002/sea2.12040.

Luxemburg, Rosa. 2003[1913]. *The accumulation of capital.* London, England and New York, NY: Routledge.

Malm, Andreas. 2013. China as chimney of the world: The fossil capital hypothesis. *Organization environment* 25(2): 146-177. https://doi.org/10.1177/1086026612449338.

Malm, Andreas. 2016a. *Fossil capital: The rise of steam power and the roots of global warming.* London, England: Verso.

Malm, Andreas. 2016b. Who lit this fire? Approaching the history of the fossil economy. *Critical Historical Studies* 3(2): 215–248.

Malm, Andreas. 2018a. Long waves of fossil development: Periodizing energy and capital. *Mediations* 31(2): 17–40.

Malm, Andreas. 2018b. *The progress of this storm: Nature and society in a warming world.* London, England: Verso.

Malm, Andreas, and Alf Hornborg. 2014. The geology of mankind? A critique of the Anthropocene narrative. *The Anthropocene Review* 1(1): 62–69. https://doi.org/10.1177/2053019613516291.

Markard, Jochen, Rob Raven, and Bernhard Truffer. 2012. Sustainability transitions: An emerging field of research and its prospects. *Research Policy* 41: 955–967. https://doi.org/10.1016/j.respol.2012.02.013.

Marks, Robert B. 2015. *The origins of the modern world: A global and environmental narrative from the fifteenth to the twenty-first century.* 4th ed. Lanham, MD: Rowman and Littlefield.

Marx, Karl. 1990[1867]. *Capital: A critique of political economy.* London, England: Penguin.

Marx, Karl, and Frederick Engels. 1969[1848]. Manifesto of the communist party. In *Marx/Engels selected works*, vol. 1, edited by Eric Hobsbawn, John Hoffman, Nicholas Jacobs, Monty Johnstone, Martin Milligan, Jeff Skelley, Ernst Wangermann, Louis Diskin, et al., 98–137. Moscow, Russia: Progress Publishers.

McGlade, Christophe, and Paul Ekins. 2015. The geographical distribution of fossil fuels unused when limiting global warming to 2°C. *Nature* 517(7533): 187–190. https://doi.org/10.1038/nature14016.

Miller, Ian, Paul Warde, Ariane Tanner, J. R. McNeill, Victor Seow, Conevery Bolton Valencius, and Robert D. Lifset. 2019. Forum: The environmental history of energy transitions. *Environmental History* 24: 463–533.

Mirowski, Philip. 1989. *More heat than light: Economics as social physics: Physics as nature's economics.* Cambridge and New York, NY: Cambridge University Press.

Mitchell, Timothy. 2011. *Carbon democracy: Political power in the age of oil.* London, England and New York, NY: Verso.

Moore, Jason. 2003. The modern world-system as environmental history? Ecology and the rise of capitalism. *Theory and Society* 32: 307–377. https://doi.org/10.1023/A:1024404620759.

Moore, Jason. 2007. Silver, ecology, and the origins of the modern world, 1450–1640. In *Rethinking environmental history: World-system history and global environmental change*, edited by Alf Hornborg, J. R. McNeill, and Martinez-Alier J., 123–142. Lanham, MD: AltaMira Press.

Nduagu, Experience I., and Ian D. Gates. 2015. Unconventional heavy oil growth and global greenhouse gas emissions. *Environment Science Technology* 49(14): 8824–8832. https://doi.org/10.1021/acs.est.5b01913.

Nikiforuk, Andrew. 2012. *The energy of slaves: Oil and the new servitude.* Vancouver, BC, Canada and Berkeley, CA: Greystone Books.

Oxfam. 2017. *An Economy for the 99%.* Oxfam briefing paper. Oxford, England: Oxfam House.

Parson, Sean, and Emily Ray. 2016. Sustainable colonization: Tar sands as resource colonialism. *Capitalism Nature Socialism* 29(3): 68–86. https://doi.org/10.1080/10455752.2016.1268187.

Pérez-Rincón, Mario Alejandro. 2006. Colombian international trade from a physical perspective: Towards an ecological "Prebisch thesis." *Ecological Economics* 59(4): 519–529. https://doi.org/10.1016/j.ecolecon.2005.11.013.

Phillips, Leigh. 2015. *Austerity ecology and the collapse-porn addicts: A defence of growth, progress, industry and stuff.* Winchester, England and Washington, DC: Zero Books.

Pinker, Steven. 2018. *Enlightenment now: The case for reason, science, humanism, and progress.* London, England: Penguin.

Podobnik, Bruce. 2005. *Global energy shifts: Fostering sustainability in a turbulent age.* Philadelphia, PA: Temple University Press.

Pomeranz, Kenneth. 2000. *The great divergence: China, Europe, and the making of the modern world economy.* Princeton and Oxford, England: Princeton University Press.

Rogelj, Joeri, Alexander Popp, Ketherine V. Calvin, Gunnar Luderer, Johannes Emmerling, David Gernaat, Shinichiro Fujimori, et al. 2018. Scenarios towards limiting global-mean temperature increase below 1.5°C. *Nature Climate Change* 8: 325–332. https://doi.org/10.1038/s41558-018-0091-3.

Rosling, Hans. 2018. *Factfulness: Ten reasons we're wrong about the world—and why things are better than you think.* New York, NY: Flatiron.

Scheidel, Arnim, and Alevgul H. Sorman. 2012. Energy transitions and the global land rush: Ultimate drivers and persistent consequences. *Global Environmental Change* 22: 588–585. https://doi.org/10.1016/j.gloenvcha.2011.12.005.

Schwartzman, David. 1996. Solar communism. *Science and society* 60(3): 307-331. https://www.jstor.org/stable/40403574.

Schwartzman, David. 2012. A critique of degrowth and its politics. *Capitalism Nature Socialism* 23(1): 119–215. https://doi.org/10.1080/10455752.2011.648848.

Sieferle, Rolf P. 2001. *The subterranean forest: Energy systems and the Industrial Revolution.* Cambridge, England: White Horse Press.

Smil, Vaclav. 2001. *Enriching the earth: Fritz Haber, Carl Bosch, and the transformation of world food production.* Cambridge, MA and London, England: The MIT Press.

Smil, Vaclav. 2015. *Power density: A key to understanding energy sources and uses.* Cambridge, MA: MIT Press.

Smil, Vaclav. 2016a. *Still the Iron Age: Iron and steel in the modern world.* Oxford, England and Cambridge, MA: Butterworth-Heinemann.

Smil, Vaclav. 2016b. *Energy and civilization: A history.* Cambridge, MA and London, England: The MIT Press.

Smil, Vaclav. 2016c. Examining energy transitions: A dozen insights based on performance. *Energy Research and Social Science* 22: 194–197. https://doi.org/10.1016/j.erss.2016.08.017.

Sovacool, Benjamin K. 2016. The history and politics of energy transitions: Comparing contested views and finding common ground. WIDER working paper, no. 2016/81.

Spash, Clive, ed. 2017. *Routledge handbook of ecological economics: Nature and society.* London, England and New York, NY: Routledge.

Steffen, Will, Jacques Grindevald, Paul Crutzen, and John McNeill. 2011. The Anthropocene: Conceptual and historical perspectives. *Philosophical Transactions of the Royal Society* 369: 842–867. https://doi.org/10.1098/rsta.2010.0327.

Steffen, Will, Johan Rockström, Katherine Richardson, Timothy M. Lenton, Carl Folke, Diana Liverman, Colin P. Summerhayes, et al. 2018. Trajectories of the Earth system in the Anthropocene. *PNAS* 115(33): 8252–8259. https://doi.org/10.1073/pnas.1810141115.

Steffen, Will, Katherine Richardson, Johan Rockström, Sarah E. Cornell, Ingo Fetzer, Elena M. Bennett, Reinette Biggs, et al. 2015a. Planetary boundaries: Guiding human development on a changing planet. *Science* 347(6223): 1259855. https://doi.org/10.1126/science.1259855.

Steffen, Will, Wendy Broadgate, Lisa Deutsch, Owen Gaffney, and Cornelia Ludwig. 2015b. The trajectory of the Anthropocene: The great acceleration. *The Anthropocene Review* 2(1): 81–98. https://doi.org/10.1177/2053019614564785.

Stivers, Richard. 1999. *Technology as magic: The triumph of the irrational.* New York, NY: Continuum.

Uddenfeldt, Therese. 2016. *Gratislunchen.* Stockholm, Sweden: Albert Bonniers Förlag.

UNDP. 2016. *UNDP support to the implementation of sustainable development goal 7: Affordable and clean energy.* New York, NY: UNDP.

UNEP. 2016. *Global material flows and resource productivity.* Assessment report for the UNEP international resource panel.

Vadén, Tere, and Antti Salminen. 2018. Ethics, nafthism, and the fossil subject. *Beyond Anthropocentrism* 6(1): 33–48.

Wallerstein, Immanuel. 2011. *The modern world system I: Capitalist agriculture and the origins of the European world-economy in the sixteenth century.* Berkeley, Los Angeles, and London, England: University of California Press.

Weisdorf, Jacob L. 2005. From foraging to farming: Explaining the Neolithic Revolution. *Journal of Economic Surveys* 19(4): 561–586. https://doi.org/10.1111/j.0950-0804.2005.00259.x.

Weisz, Helga, Marina Fischer-Kowalski, Clemens M. Grünbühel, Helmut Haberl, Fridolin Krausmann, and Verena Winiwarter. 2001. Global environmental change and historical transitions. *The European Journal of Social Science Research* 14(2): 117–142. https://doi.org/10.1080/13511610123508.

Wilkinson, Richard G. 1973. *Poverty and progress: An ecological model of economic development.* London: Methuen.

Wilson, Sheena, Adam Carlson, and Imre Szeman, eds. 2017. *Petrocultures: Oil, politics, culture.* Montreal, Quebec, Canada and Kingston, Ontario, Canada: McGill-Queen's University Press.

Wolf, Eric R. 2010[1982]. *Europe and the people without history.* Berkeley, Los Angeles and London, England: University of California Press.

Wrigley, Edward A. 1988. *Continuity, change and chance: The character of the Industrial Revolution in England.* Cambridge: Cambridge University Press.

WWF, 2018. *Living Planet Report 2018: Aiming Higher.* WWF, Gland, Switzerland.

York, Richard, and Philip Mancus. 2009. Critical human ecology: Historical materialism and natural laws. *Sociological Theory* 27(2): 122–149. https://doi.org/10.1111/j.1467-9558.2009.01340.x.

Zinn, Howard. 1995. A people's history of the United States 1492—present. New York, NY: Harper.

4   Global asymmetries in the rise of solar power[1]

Solar PV technology is based on the photovoltaic effect.[2] The effect has been known since French physicist Edmond Becquerel discovered it in his father's laboratory in 1839. By the beginning of the 20th century, Albert Einstein was awarded the Nobel Prize for discovering the laws of the photoelectric effect and the quantum theory of light that proved important for the advent of modern solar PV cells. Still, it took a few more decades before the solar PV cell was applied. The first practical PV cell was developed by scientists in Bell Labs in New Jersey, more than a hundred years after the first discovery of the PV effect. This development was demonstrated in 1954 and utilized only a few years later to power the US satellite Explorer 6. In the late 20th century, PV cells proved usable in a number of contexts, ranging from powering pocket calculators to space satellites. While this invited the aspiration to develop an advanced post-industrial society propelled entirely by renewable energy from the sun, scaling up solar PV technology remained an elusive project of the future.

It was not until the oil crisis of the 1970s that public discourse in industrialized nations entertained the notion of scaling up renewable energy technologies. Solar technology represented an alternative to fossil fuels, which were increasingly perceived as unreliable for long-term societal stability. Energy security was a prominent argument for renewable energy at the time, but a number of visionaries pointed out that renewable energy technologies were superior to fossil fuels both on social and environmental grounds (for an overview, see Mittlefehldt, 2018). Environmentalists such as Murray Bookchin (1964[2009]) and Barry Commoner (1979) argued that solar technology would decentralize power and end poverty.

The perhaps most influential spokesperson for renewables (at least in the US) was Amory Lovins, who in 1976 introduced the idea of the "soft energy path" (Lovins, 1976, 1979). Lovins argued for a society relying on energy captured directly from the sun and the wind. Such a society, he argued, would at once be more efficient and provide a technological basis for reduced resource consumption. Lovins was directly influenced by the work of Ernst Schumacher who in *Small is Beautiful* argued for "intermediate technology" (what was later called "appropriate technology"). Schumacher defined this famous concept as follows:

DOI: 10.4324/9781003292319-4

I have named it *intermediate technology* to signify that it is vastly superior to the primitive technology of bygone ages but at the same time much simpler, cheaper, and freer than the super-technology of the rich. One can also call it self-help technology, or democratic or people's technology - a technology to which everybody can gain admittance and which is not reserved to those already rich and powerful (Schumacher, 1993[1973]: 127, emphasis in original).

In comparison to nuclear power, proponents saw solar technology as an appropriate technology, at once more democratic and environmentally friendly. With a growing awareness of anthropogenic climate change in the 1990s, the argument for renewable energy technologies shifted in focus (Hake et al., 2015). Societies could now reduce the dangerous greenhouse emissions by installing solar technology. So the discourse on renewable energy was reinvigorated. Herman Scheer, an influential German visionary and green-wing politician, laid out a vision for a future society sustaining on energy harnessed directly in local environments that was at once more efficient, community empowering, and environmentally friendly (Scheer, 2005, 2007). Today, the conventional attitude and feeling is that solar PV technology still is an appropriate, democratic, environmentally friendly, and empowering technological option (see Chapter 1).

It is important to recognize that these visions and arguments for solar power were articulated at a time when manufacturing and installations of solar PV modules remained modest and small-scale. As expressed in an article from *Science Magazine* a few years ago, "the recent rapid declines in PV system pricing illustrate that we are entering an era in which PV already is or will soon become cost-competitive with conventional electricity generation in many parts of the world" (Haegel et al., 2019). Over the last 20 years, the installation of solar PV capacity has soared beyond most expectations and solar PV modules are rapidly becoming commodities available in bulk for richer societies in the world economy.

In this chapter, I explore how the global commercialization of solar PV technology problematizes the sentiment that solar PV technology is a democratic and environmentally friendly technology. If Schumacher was a proponent of "appropriate technology," he was also an ardent critic of "technology of mass production," which he saw as "inherently violent, ecologically damaging, self-defeating in terms of non-renewable resources, and stultifying for the human person" (1993[1973]: 127). Recent developments have led to the ironic historical moment in which some "appropriate technologies," notably solar PV modules, have now become "technologies of mass production." While the biophysical condition of solar PV technologies has changed with the introduction of mass manufacturing (as we will soon explore), the inherited techno-political visions of the 1970s have essentially remained unchanged. As exemplified by the case of Katrineholm (see Chapter 1), visions of a socially just and ecologically sustainable future based on solar PV technology have not fully integrated the new reality of PV mass manufacturing.

In what follows, I give a brief account of the Chinese PV manufacturing boom and a review of the most common explanations for the boom and show how they rest on the assumption that technology is immaterial. This is followed by an empirical investigation into the presence of ecologically unequal exchange between Germany and China during the time of the boom. (The reason why I focus on these two countries will soon become apparent.) I then discuss the implications of these results and discuss how the solar boom in solar PV manufacturing and trade was contingent on uneven flows of embodied labor and embodied resources in the world economy. Given the results of this investigation, I conclude that societies propelled by solar PV technology are not by default democratic and sustainable and that ecologically unequal exchange may be inherent to large-scale solar PV development. This forces us to nuance the notion of solar PV technology as an inherently "appropriate technology" for global environmental justice.

The solar boom

Over the last two decades, the global solar photovoltaic market reshaped the meaning of solar PV technology. The most apparent consequences of this development have been that prices of crystalline solar PV modules plunged, while global installations of PV systems soared. Between the years 2000 and 2012, the price per watt peak (Wp) manufacture fell from 3.7 to 1 USD, only to plunge to 0.25 USD/Wp in 2018. Meanwhile, cumulative global installations of PV capacity soared from a modest 1.2 GW in 2000 to a momentous 500 GW in 2018. The added capacity in the year 2018 alone was equivalent to the world's historical cumulative capacity up until 2012 (Figure 4.1). This period of industrial frenzy has aptly been called a "boom" in the global PV market. Judging from forecasting scenarios, the manufacture and installation of PV technologies is believed to increase over the two coming decades, with installations possibly doubling in the coming few years (Jäger-Waldau, 2019; UNEP, 2020).

There is little doubt that the boom in the global solar PV industry correlates with the entrance of China into the international market as a manufacturing and exporting nation, motivated and encouraged by European demand (Quitzow, 2015; Yu et al., 2016; Nemet, 2019). Dustin Mulvaney, in his book *Solar Power: Innovation, Sustainability, and Environmental Justice*, considers the emergence of the entrance of China as "arguably the most significant story in renewable energy over the past decade" (2019: 224). Even if Chinese firms have only been engaged in industrial solar PV production since the mid-1980s, the significance of Chinese industry cannot be underestimated. Compared to US industries that pioneered the development of the solar PV cell in the 1950s, Chinese actors were latecomers to the PV industry. While the Chinese manufacture of crystalline solar cells increased considerably during the 1990s, the Chinese PV market was then almost exclusively geared toward small off-grid rural electrification projects that did not facilitate much industrial growth (Zhang and He, 2013: 394).

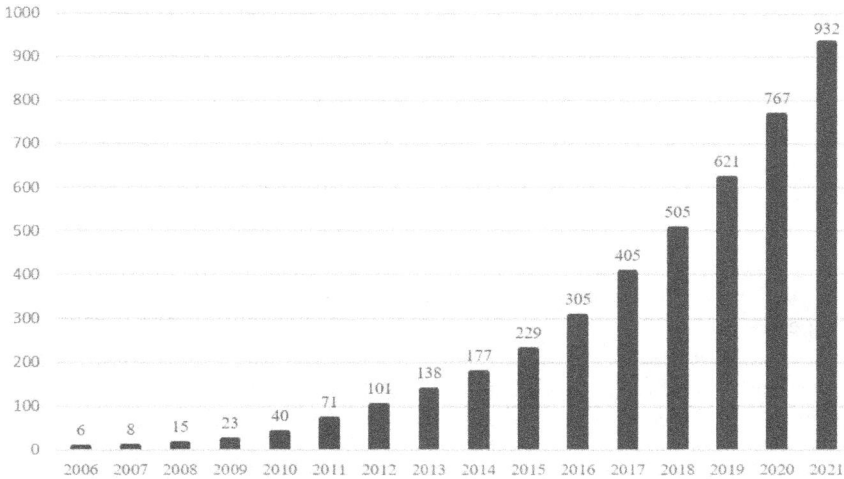

Figure 4.1 World total installations of solar photovoltaic capacity in 2006–2021 (GW). Data source REN21 (2016, 2019, 2022). Throughout this book, I rely on easy-access data from potentially problematic sources, such as REN and BP. Energy indicators from such sources are sometimes intentionally designed for green growth narratives and unsustainable political targets (Rodriguez et al., 2020). I ask the reader to keep this in mind even if I draw critical conclusions from this data, which are contradictory to the political interest of these sources.

This situation gradually changed with the growing international demand for solar PV technologies driven by a progressive German energy politics, starting with the red-green coalition formed in 1998. As part of the long red-green effort to phase out nuclear power and transition to a renewable energy mix, the coalition offered generous feed-in tariffs (FITs) supported by the Renewable Energy Act (EEG) in 2000. This resulted in a remarkable increase in global solar PV demand and installation (Hake et al., 2015). Despite amendments to the act in 2009 to reduce FITs, German demand continued to soar and soon Germany was the largest PV market in the world, representing 50% of world PV module demand. With this demand, there was a need for a large increase in production that German solar PV corporations could not meet (despite fiscal support). This represented an opportunity for Chinese entrepreneurs to start up, or shift, to production for the global solar PV industry to turn considerable profits (Nemet, 2019). Soon, German equipment providers selling manufacturing equipment established links with Chinese PV manufacturers exporting solar PV modules back to Germany (Quitzow, 2015).

As noted in an article in *Scientific American*, "Germany … provided the capital, technology and experts to lure China into making solar panels to meet the

German demand," and "'the Chinese took it'" (Fialka, 2016). Encouraged by European (mostly German, but also Spanish) demand and armed with low-cost labor, cheap coal-propelled manufacturing, and generous government loans, Chinese companies initiated the "explosive increase" in global solar PV module manufacturing by the mid-2000s (Yu et al., 2016: 466). China quickly rose as the largest solar PV manufacturer in the world, producing nearly 30% of all solar PV modules in 2007. In 2011, this figure had increased to 60% and in 2018 to 64% (Jäger-Waldau, 2019: 2). By 2017, 95% of the world production of solar PV modules was located in Asia (Mulvaney, 2019: 224). Meanwhile, "China's solar-electric panel industry dropped world prices by 80 percent" (Fialka, 2016). Between the years 2008 and 2009, after a period of polysilicon shortage, the price on silicon PV cells dropped by almost 50% (Figure 4.2).

As of 2006, a whole 96% of China's PV production was geared toward the international market (i.e., exported; Yu et al., 2016: 466). However, as international demand diminished due to trade regulations and the financial crisis of 2008, the Chinese government responded in its 11th five-year plan (2006–2010) by increasing domestic demand in a series of national PV-financing programs, such as the "Golden Sun Demonstration Program." Following these plans, China also became the largest *consumer* of solar PV technologies by 2015, even surpassing Germany as the previous leader in solar PV installations (REN21, 2019). Today, China is unrivaled in both the production and installation of solar PV modules, even if at least half of the modules manufactured in China are still produced for export.

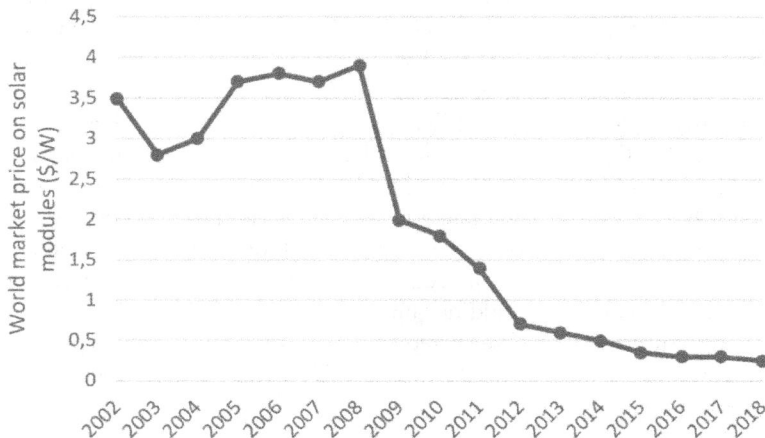

Figure 4.2 Average selling prices of solar PV modules, 2002–2018. Data from Nemet (2019) and Haegel et al. (2019). Figures 4.2-4.9 originally appear in the paper published by the author in *Ecological Economics*, 199 under the CC BY license (Roos, 2022).

The effects of the boom were systemic. It upset international relations between China and other nations in the world, as Chinese solar manufacturing companies effectively shattered all competition and glutted the global market with cheap products. The four largest solar manufacturers in Germany (Q-Cells, Solar Millennium, Solon, and Solarhybrid) and no less than 16 US solar companies (including BP Solar, Evergreen Solar, SolarWorld, and SolarPower) closed down, off-shored production to Asia or were sold as a result of the rise of Chinese manufacturers. In Japan, a staggering 65 solar-related companies filed for bankruptcy and had to close down. The German government, who was offering generous FITs as incentive for increased rooftop installations, was threatened with heavy financial costs as installations soared. In an effort to defend the national economy and German manufacturing profits and jobs (and by extension, the German consumer base), it decided to reduce funding in its FIT program. Later, the European Commission levied import tariffs and anti-dumping fees on a majority of cheap solar PV technologies imported from China. This was met by fierce opposition from some European solar corporations and Environmental Non-Governmental Organizations (ENGOs) such as the World Wildlife Fund (WWF) and Greenpeace, who saw it as an obstacle to a rapid transition to renewable energy (Reuters, 2017).

In the US, some solar corporations successfully lobbied for import tariffs on Chinese solar PV modules to protect their businesses. This resulted in Chinese PV products being subjected to tariff fees of up to 250%. When these measures were implemented, it sparked a "trade war" between China and the US, in which each accused the other of anti-free trade measures. In part to avoid European and US tariffs and in part due to other factors contributing to increasing production costs (e.g., labor prices), large Chinese solar companies such as Trina Solar relocated their factories to Southeast Asian countries including Thailand and Vietnam (Roselund, 2016; Sinn, 2019; Watt et al., 2019).

Another effect of the boom was the noticeable increase in demand for materials needed in the manufacturing of solar PV modules. In terms of materials, the mono- and poly-crystalline PV technologies that make up around 90% of the global PV market requires solar-grade silicon, silver, lead, and nickel (Grau et al., 2011: 3; WB, 2017). In comparison, less commercially viable technologies such as thin-film technologies require more specific material compounds, including tellurium, indium, cadmium, and zinc, which are often higher in toxicity (Exter et al., 2018; Muteri et al., 2020). Much attention has been paid specifically to silver, which is a potentially scarce material for a large-scale development of the now commercially viable PV technologies (Piano and Mayumi, 2017). One recent study has shown that the booming production of solar PV technologies caused a definite increase in global silver prices (Apergis and Apergis, 2019). Similarly, solar-grade polysilicon production boomed with the rapidly expanding solar PV industry and China's initiative to create a domestic polysilicon production industry (Bernreuter, 2020).

The sheer amount of materials mined and processed for the rapidly expanding solar PV industry has become a basis for social and environmental concern. Silver has been found in high concentrations in various Chinese municipalities (Chen et al., 2020), where it has shown to be a driver of eutrophication and terrestrial eco-toxicity (Muteri et al., 2020: 17). Polysilicon production has also been pointed out as hazardous for workers and environments due to dangerous chemicals such as silicone tetrachloride and hydrofluoric acid leaking into environments. In an article in *Nature*, Yang et al. (2014) reported that Greenpeace and the Chinese Renewable Energy Industries Association found that two-thirds of all solar companies in China failed to meet the national environmental regulations. This was reported even after the scandal of 2011, when rest-products from a solar-panel factory in Haining City were discovered to have severely polluted the Mujiaqiao River. The company, Jinko Solar, had been dumping toxic wastewater into the city river that killed a considerable amount of fish. This was met by a local uprising against the factory that was suppressed only after three days by the Chinese police's "heavy-handed tactics" (Chan, 2011).

In addition to this, there are several environmental, health, and safety hazards throughout the production chain of solar PV technologies that are less regulated in Chinese factories (Wang and Feng, 2014). These hazards allow for production at lower prices. LCA analyses frequently find that low-cost production in Chinese PV industries is associated with higher (up to double) social-ecological impacts in comparison to US or European industries (Yue et al., 2014; Stamford and Azapagic, 2018). These include, most notably, much higher energy consumption and greenhouse gas emissions. It is no secret that China's rapid ascent as a global economic super-power is linked to its low-cost labor force and massive burning of coal to propel its industries (Malm, 2013). Consequently, as PV production companies relocated from Europe and the US, the carbon footprint of solar PV modules increased drastically.

In the case of the boom in solar PV production, some have argued that relocating the production of solar PV modules to China resulted in "energy cannibalism" in which the energy and emissions payback has been offset by aggregate growth in annual solar PV module production (Pearce, 2009; Decker, 2015). The annual amount of energy dissipated and greenhouse gases emitted in the production of solar PV modules is canceling out the annual mitigating potential embodied in the installed PV technologies. Taking into consideration factors such as solar insolation, geographic distribution, and LCA accounts, Decker (2015) arrives at the conclusion that the maximum sustainable growth rate for the now global solar industry is 16% per year. Based on these figures, one important effect of the boom is that the manufacturing and installation of solar PV technologies may itself have facilitated – not mitigated – increased energy consumption and greenhouse gas emissions.

In light of these social-ecological conditions, recent research has begun to seek ways for mitigating environmental injustices throughout the production chains of solar PV technologies (e.g., Mulvaney, 2019). These, however, typically do not focus on understanding whether the observed social and environmental problems were necessary conditions for meeting the German demand and the low manufacturing costs. Even Decker (2015), who provides a highly critical analysis, does not consider how the price differences between US/European nations (notably Germany) and China may have been a necessary condition for it to occur in the first place. While we should seek to "ensure that solar energy commodity chains evolve in a just and sustainable way" (Mulvaney, 2019: 248) and "carefully select locations for production and installation" (Decker, 2015), we must ask whether such interventions would impede the expansion of solar PV technologies.

This means that we should ask what characterized the international relation between Germany and China over the course of the boom in solar PV technology. Why, in other words, could Germany not supply its own solar PV technologies? Why did Germany outsource the manufacture of its energy-political vision? If Germany possessed the "capital, technology and the experts," then why could German companies not employ these assets to ramp up its own domestic production? In this chapter, I argue that the world division of labor between Germany and China – characterized by an ecologically unequal exchange – may have been a key condition for the solar boom. However, before we consider this interpretation, we shall look at some of the already existing explanations for the boom and their underlying philosophical assumptions.

Immaterial explanations for the boom

The present explanations for the boom all understand solar PV technology as an ontologically immaterial phenomenon. The solar PV boom and the plunging prices are, according to these explanations, supposed to have sprung forth from ideas. These explanations can be found among a variety of sources represented by both academics and non-academics. The following explanations are not exhaustive, but they provide a general overview of the conventional interpretations of the boom.

Instrumental explanations: the boom originated from scientific knowledge

The first explanation is the conventional notion that breakthroughs in scientific knowledge and engineering design contributed to the plunging prices in solar PV technology. This explanation centers on the notion that efficiency improvements push down prices per unit produced by allowing more to be done with less. This explanation focusing on efficiency takes at least two forms. The first form emphasizes how improved engineering design of the technological artifact improves efficiencies and thereby reduces prices. Examples of this include findings in a study from MIT showing that increases in module efficiency may have contributed up to 25% of the decline in prices between 1980 and 2012 (Kavlak et al., 2018). However, the same study pointed out that the increase in solar power plant size and policies likely had larger effects on the drop in prices from 2001 onwards.

The second form emphasizes the increases in efficiency in the wider sphere of installation and manufacturing that comes from applied, or practical, knowledge over time (e.g., Haegel et al., 2019). The underlying assumption here is that the more frequently something is done, the more efficiently and cheaper it can be done. This is referred to as the "experience curve" (or Henderson curve) explaining and modeling how increases in production correlate with more efficient practices over time. Aaron Bastani, in his *Fully Automated Luxury Communism*, falls back on this model in explaining the plunge in solar PV module prices. Here it is worth quoting Bastani (2019: 47–48) at length.

> [T]he most important area where one sees the experience curve at work is with the price of photovoltaic (PV) cells. … Here progress correlates almost perfectly to what Henderson would have predicted, with the cost of PV falling 20 per cent every time capacity doubled over the last sixty years. When the technology was deployed for the first time aboard NASA's Vanguard 1 satellite in 1958, each panel was able to generate a maximum half a watt of energy at a cost of many thousands of dollars each. By the mid-1970s, that figure had fallen dramatically to $100 per watt … Yet by 2016 the price-performance ratio of solar had been transformed, with a watt of energy from solar array costing as little as fifty cents … *and with global solar capacity doubling every two years … a virtuous cycle between increased capacity and ever-falling prices has been established* (emphasis added).

Here we can see how the knowledge derived from larger amounts of PV modules installed is itself understood as a driver for the increased PV module demand. In other words, lower prices driven by increased knowledge are creating an increase in demand for solar PV technologies (i.e., knowledge → efficiency → lower prices → demand → installations → knowledge). However, as we have seen in the case of the boom in the global PV market, the demand was clearly driven by German political interests, not by progress in PV engineering design. Other than reiterating technological determinism, where knowledge via technological progress is believed to determine social change, this explanation fails to explain why German companies, who possessed expert knowledge in solar PV manufacturing, were outcompeted by Chinese entrepreneurs who had previously produced water purification products, cosmetics (Suntech), or household detergents (Trina Solar) (Nemet, 2019: 148–150).

This is not to say that knowledge and designs that are more efficient did not have any effect on PV module prices. But the claim that increased experience in the wider sphere of manufacturing and installation were key mechanisms driving the boom contradicts the actual events. To be sure, ideas surrounding solar PV technology served a discursive function for German red-green politics, but this had been the case already since the early 1970s (Hake et al., 2015). What changed just prior to the boom was not any spectacular increase in knowledge or technological breakthrough in cell efficiency, but a shift in power in favor of red-green politics that advocated solar PV technologies increasingly manufactured at cheap prices in Asia.

Chinese governance: the boom originated from Chinese governance

The second explanation focuses on how the powerful central state of China strategically planned and implemented measures to develop its PV manufacturing industry. This explanation stems from considerations on how China, sometimes described as a "developmental state" or as an "environmental authoritarian" state, implemented top-down measures to support PV manufacturing companies located within its borders (Gilley, 2012; Chen and Lees, 2016). This explanation bears some resemblance to Feenberg's critical theory of technology, which highlights the importance of different social forms in designing and developing particular technologies (see Chapter 2). However, rather than being concerned with the design of a specific technological artifact, this explanation for the global boom in PV installation focuses on how the Chinese government, operating with social-political interest disparate from Western nations driven by corporate profit-maximization provided a unique environment for solar PV manufacturing to flourish.

Writing for *Aljazeera*, Larry Beinhart (2018) argues that it was Chinese development policies and planning that catalyzed the rise of PV manufacturers in China, which ultimately led to the global commercialization of solar PV technology. These policy designs, modeled from South Korea, include the nationalization of commercial banks and state control over credit, which allowed its "solar industry to prosper as part of a long-term, well-thought-out industrial development strategy" (ibid.). Beinhart points out that in comparison to China, the US were "against... an articulated, thoughtful, deliberate industrial policy." This, Beinhart points out, explains why China and not the US is now the global leader in the solar PV market. As with the case of German demand, this explanation emphasizes how policies, driven by different forms of governance with specific social-political interests, influence the prospects for solar PV development.

The notion that the Chinese government supported solar PV manufacturers from a position of a "well-thought-out industrial development strategy" can be questioned on the basis that the Chinese government generally lacked an interest in PV manufacturing until Chinese firms appeared in Western stock exchange around 2008 (Nemet, 2019: 138–139). Notably, the interest of the Chinese government was kindled only after Chinese manufacturers produced a majority of the world's PV modules (Yu et al., 2016: 466). The Chinese government, it seems, was reactive in the development of the solar PV manufacturing industry that was crucial to meet German demand. The Chinese Renewable Energy Law of 2005, which is commonly pointed out as a core mechanism by which the Chinese government supported the solar PV industry, was at the time geared toward increasing domestic demand and installation and did not seem to have any direct importance for manufacturers (Martinot, 2010). The law can be understood as the Chinese equivalent to the German EEG, developed with expert help from both Denmark and Germany. In addition to this, Nemet (2019: 148), who recently documented the development in his *How Solar Energy Became Cheap*, explains how pioneering Chinese corporations, such as Suntech, relied on foreign direct investment from Western corporations such as Goldman Sachs for their economic success. Only after the economically successful establishment of Chinese PV manufacturers did the Chinese government take an active interest in their continuation.

While Beinhart provides an important comparison between China and the US, this raises the question why German companies – who also benefited from generous government policies – could not scale up to meet German consumer demand (see Grau et al., 2011). Here, the economic impact of the financial crisis in 2008 may have played a larger role than differences in governance. In Germany and other European nations, government support (such as FITs) had to be cut down as economies went into recession (Yu et al., 2016). In contrast, the Chinese economy was not affected in the same way by the crisis, in part due to its strength in the world economy and in part due to previous experiences with financial crises (Li et al., 2012). Consequently, even at the height of the crisis in Europe in 2008, China's government had the capacity to strengthen policies for increasing domestic PV installation. At this time, Chinese development banks also offered generous loans for solar companies struggling to turn a profit (Dong et al., 2015a). The degree to which the role of China in the world economy was strong due to its particular form of governance (state capitalism) is debatable. It would seem, however, that the different world-economic positions of Germany and China provide a better explanation for why the development of a world-scale solar PV manufacturing capacity occurred in China and not in Germany (as I discuss below).

International co-evolution: The boom originated from knowledge exchange

The third explanation focuses on how the global solar PV market emerged from a complex international network of entrepreneurs, corporations, and government actors engaged in knowledge exchange. This explanatory framework acknowledges that China's capacity to scale PV manufacturing was an important reason why a commercially viable solar PV industry could emerge. While it is sometimes acknowledged that prices in material input factors, such as wages, rent, energy, and materials, had an initial significance for the Chinese capacity to meet German demand, this framework considers "knowledge" (sometimes referred to as innovation or expertise) as the supreme input factor shaping the dynamics of the global PV market. Curiously, even the Chinese capacity for mass manufacturing is here understood as something originating from knowledge as opposed to originating from specific material conditions characteristic of the Chinese economy, such as the heavy reliance on coal.

In contrast to the purely instrumental explanation above, the knowledge in question is not understood to originate from scientific expertise or engineering design, but from actors making up an international network of knowledge exchange. In other words, the boom did not occur because Chinese actors were unique in their ability to innovate or scale up manufacturing but because an international network of knowledge exchange emerged in which European, Australian, and US actors could trade knowledge with Chinese manufacturers and vice versa. Helveston and Nahm (2019: 395) conclude, for example, that

> Chinese manufacturers gain technological know-how from advanced foreign incumbents, and the foreign partners feed the manufacturing and scale-up solutions their Chinese partners identify back into up-stream R&D activities.

In this explanation, knowledge exchange typically constitutes a win-win situation where each actor or nation has something to benefit from engaging in knowledge transfer (sometimes called "technology transfer") to optimize or improve their respective manufacturing capabilities. This means that there was no unique event, condition, or knowledge within China or among Chinese corporations that drove the boom in global solar PV manufacturing. Certainly, the Chinese capacity to scale was important, but the emerging PV industry, as put by Quitzow (2015: 143), was "a co-evolutionary process of mutual cumulative causation," where the knowledge exchange between Chinese and German actors co-created the boom.

Importantly, these studies acknowledge that the co-evolution underlying the emerging PV industry did not necessarily imply that the exchanges were always equal or just. As pointed out by Nahm and Steinfeld (2014), solar PV products that are "made in China" typically benefit US and European firms who, unlike their Chinese partners, do not have to navigate substantial financial or environmental risks. In this explanatory framework, a world division of labor is sometimes mentioned in passing, but quickly downplayed with reference to manufacturing automation or the rapid increase in Chinese technical know-how. For example, Helveston and Nahm (2019: 796) argue in their concluding remarks that even if world asymmetries are present today, "[t]he division of labor between Western inventors and Chinese manufacturers is not fixed or inevitable." This is true, but it should not deter us from questioning to what degree a world division of labor was a necessary condition for solar PV commercialization, and to what degree it may be a permanent (albeit geographically mobile and flexible) condition for modern technological progress in general.

That is to ask, could the boom have occurred without this particular division of labor? If yes, then, why did it not? Helveston and Nahm (ibid.) continue by saying that, "in a global marketplace such as energy technology, it is unlikely that the entire value chain for a complex, manufactured product would lie entirely within national boundaries." Again, this is certainly true, but it should not deter us from questioning why certain parts of the supply chain – typically primary sectors – are systematically placed in lower-wage, peripheral or semi-peripheral regions of the world. To turn a blind eye to the world division of labor under which the solar PV boom occurred is arguably to miss the most important historical condition under which solar PV technologies were commercialized.

By focusing on how *efficiency improvements, governance,* or *knowledge exchange* creates conditions for increased production, conventional explanations of the solar boom omit the role of particular human-environmental relations and the energy and material flows that make up the physical production of solar PV technologies. This is especially peculiar if we consider the vast amount of evidence pointing at a strong correlation between mass production (i.e., industrialism) and the access to fossil fuels (see Chapter 3).

Ecologically unequal exchange between Germany and China

The relation between Germany and China had a particular significance for the boom in production of solar PV modules. As some studies have already suggested, it makes little sense to understand China and Germany as isolated countries in this historical context (see the section above). Rather, it may be best to understand that these countries, along with their associated actors, make up a particular relation that itself catalyzed the boom. Paradoxically, to understand such a relation, we must understand some basic qualitative differences between the countries that allow for such a relation to emerge.[3] This implies that we should seek to understand that each country in the modern world economy, even if unique, develops in a co-evolutionary process. Crucially, however, such a relation must be understood to constitute not only innovative knowledge or governing orders, but also biophysical flows of matter-energy. Knowledge and governance are certainly important, but neither exist apart from a transformation of matter-energy (see, e.g., Strumsky et al., 2010). Therefore, the question is what kind of material relation that emerged historically between Germany and China and what role this relation played in the boom of the global PV market. To understand this, we will first very briefly look at the history of each country and then calculate the material relations that catalyzed the boom itself. Finally, I will discuss ecologically unequal exchange as a necessity for solar PV technology becoming a commercially viable option in the 21st century.

Resource exchange between China and Germany

From a world-systems perspective, Germany is a core country and China is a semi-peripheral country. Ranked by the World Bank as a high-income economy, Germany is known for its internationally strong vehicle industry, medical industry, and the production of machinery for export. Measured in monetary exchange value, the Observatory for Economic Complexity (OEC) ranks Germany as one of the most complex and largest exporting nations in the world.

This ranking is linked to a very specific historical past. Germany was one of the first countries to transition from an agrarian to an industrial regime, with a rapid increase in the use of fossil energy during the 19th century (Fischer-Kowalski et al., 2019). Already by the 1860s, over 50% of the energy throughput in the

German economy was fossil-based (Molina and Toledo, 2014: 337). In addition to this, Germany has a colonial legacy, spanning roughly from the onset of the Scramble for Africa (1880) to the end of the First World War (1918). At the height of its power in 1914, the German Empire could draw on resources from an additional 10.6 million square kilometers and 55.5 million inhabitants outside its borders (ibid. 225). Germany's early industrialization combined with its colonial legacy has no doubt shaped the country's contemporary profile in the world economy (for the case of France, see Infante-Amate and Krausmann, 2019). Despite its turbulent 20th-century history, with economic depression, two world wars, fascism, and the cold war period, Germany is today a strong state and a top actor competing for increased economic power in the world economy.

Today, Germany boasts ambitious environmental targets and some of the most promising trajectories for 'sustainable development' according to conventional discourse (UNEP, 2011). However, the quantifiable difference between Germany's climate targets and actual environmental mitigations – the so-called "emissions gap" – remains wide (CAT, 2020b). Germany's environmental loads far exceed the carrying capacity of the German land base and globally sustainable footprints per capita (WWF, 2020). In part, the conventional narrative is kept alive because Germany is importing much of its energy-matter throughput via international trade. Several studies have documented how Germany as a core, high-income country, relies upon a net import of matter-energy in the world economy (Giljum and Eisenmenger, 2004; Jorgenson, 2009; Dorninger and Hornborg, 2015). While Germany is a net exporter in terms of monetary exchange value, the country is a net importer of embodied matter-energy. Germany's ambitious environmental politics is thereby associated with an environmental load displacement, whereby Germany exports a considerable amount of the environmental degradation resulting from its economic development to other countries. From a biophysical perspective, then, Germany exhibits a clear core-like position in the world economy.

Now we turn to China. According to the World Bank, China ranks as an "upper middle income" country. Currently, China is the largest exporting nation in the world, mainly exporting machinery (broadcasting equipment and computers), textiles, and a vast range of other commodities. China has an immensely rich cultural and material history. The region of contemporary China is associated with some of the earliest and most complex civilizations in history and at least a dozen dynasties well before the 20th century. World-system and dependency scholars Kenneth Pomeranz and Andre Gunder Frank have argued against a Eurocentric understanding of history by stressing the central role of China in the formation of the modern world economy, not the least with reference to the Chinese demand for silver as a core incentive for early European colonialism (see Chapter 3).

Despite China's central role in the early modern period, the country was late to industrialize. Several explanations exist as to why China did not industrialize earlier, or simultaneously with western European nations. Pomeranz (2000), for example, suggests that China's deteriorating relations to its peripheries contributed to halting its industrialization. To this, we could add the deteriorating social conditions in China imposed by the British Empire with the forceful opening of the Chinese markets for opium trade. The two opium wars that followed established relations between China and European colonial powers that benefited the British Empire (e.g., Clark and Foster, 2009). With impediments such as these, combined with the fall of the Chinese Empire in 1911, two world wars, and the stumbling Maoist attempt at industrialization through the Great Leap Forward, Chinese industrialization did not take off until well into the 20th century, a whole century after Germany and other western European countries (Fischer-Kowalski et al., 2019). The economic reforms of 1978 are typically pointed out as a key turnaround event that served as a new foundation for modernization and industrialization in China. Since then, China has quickly risen as the second most powerful economy in the world.

During this time of rapid economic growth, China has been fundamentally dependent on coal as the energetic basis for its expanding industries. Energy expert Vaclav Smil (2016: 275) explains that "China's total coal output more than quadrupled between 1980 (907 Mt) and 2013 (3.97 Gt), when it accounted for almost as much as the rest of the world production put together." Even if the relative amount of coal in the Chinese energy mix is slowly being reduced, Chinese industries continue to burn larger and larger amounts of coal every year to propel its increasing economic activity, which provides massive amounts of commodities to the world. Given its high fossil fuel dependence and the vast amounts of extraction of other non-renewable resources, China is often described as one of the worst polluters in the world. The *Climate action tracker*, for instance, categorizes China's efforts at decarbonization as "highly insufficient," estimating that the Chinese emissions reduction targets are in a range that would warm the Earth to a detrimental 3°C–4°C above pre-industrial levels (CAT, 2020a). In a highly globalized economy, however, it is problematic to place blame simply upon the Chinese government for this trajectory, since the country is an arena for a number of national and international actors with economic interests in producing cheap commodities for a global market (Dong et al., 2015b; Shen, 2017). Considerable parts of the environmental damages from Chinese production processes benefit other countries or actors by virtue of commodities being exported to other world regions or providing profits for non-Chinese actors.

In a study assessing the physical trade between China and 186 other countries in the world economy, it was found that developed regions (including the EU) displace their environmental loads to China through trade (Yu et al., 2014). This environmental load included embodied greenhouse gas emissions (CO_2 and SO_2) as well as embodied water and embodied land. The same study concluded that China in turn exports some of its own environmental loads to less-developed regions of the world, including Southeast Asian and African countries. China, seen from a biophysical world-systems perspective, thereby exhibits both core- and periphery-like processes, marking it as a semi-periphery in the world economy. Several other studies have confirmed the Chinese role as semi-periphery, as it simultaneously exports embodied greenhouse gas emissions, embodied energy, embodied land, and embodied material while it *imports* embodied forests, embodied land, and embodied water (Yu et al., 2014; Peng et al., 2016; Tian et al., 2017; Shandra et al., 2019).

Given their respective roles in the world economy, it is not surprising that direct trade relations between China and Germany have been shown to facilitate an ecologically unequal exchange whereby embodied matter-energy is transferred to Germany (Tian et al., 2017). It is interesting to note that the most relevant trade sectors between China and Germany, considered in physical measurements, included the German export of "machinery" to China and the Chinese export of "electrical and optical equipment" to Germany (Tian et al., 2017). This means that the German export of machinery, such as cars, PV equipment, and similar products, has a biophysical significance for the Chinese economy, while electrical equipment, such as computers and solar PV modules, have a physical significance for the German economy.

An LCA-based account of ecologically unequal exchange

This chapter now applies an LCA-based method for assessing the significance of ecologically unequal exchange in the rise of solar power. Many quantitative studies on ecologically unequal exchange focus on specific nations or world-economic regions corresponding to the world division of labor described in world-system analysis. As we have already seen how the world-economic relation between Germany and China was characterized by ecologically unequal exchange, the goal here is to calculate the presence of unequal exchange in the trade with focal commodities of the global solar PV market. Through an LCA-based method, it is possible to assess ecologically unequal exchange in trade with specific commodities (Oulu, 2015). This method gives a detailed understanding of the extent to which trade in specific commodities is involved in asymmetric flows of resources in the world economy.

The LCA-based method is suitable for calculating the asymmetric flows of embodied resources in trade with specific commodities, but this method has not been used to explain whether such asymmetric flows influence the financial or biophysical viability of the commodities traded. In theory, even if the trade in solar PV modules and solar PV manufacturing machines involves an asymmetric transfer of resources, the financial and biophysical viability of solar PV modules might still be high if other industrial sectors provide the means for a (biophysical or monetary) subsidy. I address this issue by providing an estimate of what the solar PV modules would cost if they had been produced with German wages and energy prices. As such, I assume that market prices (USD/W) influence the cost-effectiveness of solar PV development. I propose that an LCA-based assessment of ecologically unequal exchange (EUE) may be used to understand the extent to which a commodity's price – and so financial viability – is influenced by the geographical location of production.

To measure this, I introduce the notion of a "displacement-adjusted price," which denotes the price of a commodity if the displaced environmental loads (through EUE) were to be supplied with domestic wages and prices. I also introduce the notion of "displacement-based efficiency," which denotes the physical efficiency of a technology after the associated environmental load displacements have been analytically excluded as necessary inputs. For this chapter, I employ the concepts of energy return on energy investment (EROI) and power density for assessing the displacement-based efficiency of solar PV technology. By comparing the displacement-adjusted price with the market price and the displacement-based efficiency with the actual efficiency, it is possible to evaluate the significance of ecologically unequal exchange for the financial and biophysical viability of one or two commodities – in this case solar PV modules.

LCA-based calculations on ecologically unequal exchange have two parts. First is to calculate the embodied resources and impacts of the individual commodities traded, and second is to calculate the bulk exchange of these commodities in the world economy relative to a fixed market price. This includes four sub-phases:

- First, a scope definition should be provided wherein the functional unit, system boundaries, and units of measurements are articulated.
- This is followed by an inventory analysis wherein the resource intensity per functional unit is determined and presented. This chapter relies on secondary data gathered from previously published LCA studies.
- Third, an impact assessment is made wherein unequal exchange is determined via a comparison of the resource intensity per unit of exchange value.
- Finally, the implications of the results are discussed.

To compare the ecological exchange implicated in two focal commodities in the solar PV market, the functional units were defined as follows: One Chinese solar multi-crystalline silicone solar photovoltaic module and one German solar photovoltaic manufacturing machine. The system boundaries were set to include measurements traced from the extraction of the necessary matter-energy to the assembly of the final product (Figures 4.3 and 4.4). Since this chapter is focusing on the exchange of finished products, environmental impacts associated with the usage, disposal, and recycling phases are considered outside the system boundaries. Only domestic resources and greenhouse gas emissions are considered, even if the commodity chains extend internationally to a lesser degree (for solar PV, see Dong et al., 2015a, b). The units of measurement include embodied land, embodied labor, embodied energy, and embodied CO_2-eq. emissions. These units of embodiments are then related to quantities and monetary exchange values (USD) of the respective commodities, derived from the UN database COMTRADE (2021) and the Trend Economy (2020) database on commodity exchanges.

Table 4.1 summarizes the resource intensity of a Chinese solar PV module. Methodologically, the resource intensity in the life-cycle inventory table is presented as if it is static in time, which means that variations in resource intensity associated with changes in the manufacturing process have not been taken into consideration. To avoid portraying the resource intensity as larger than it

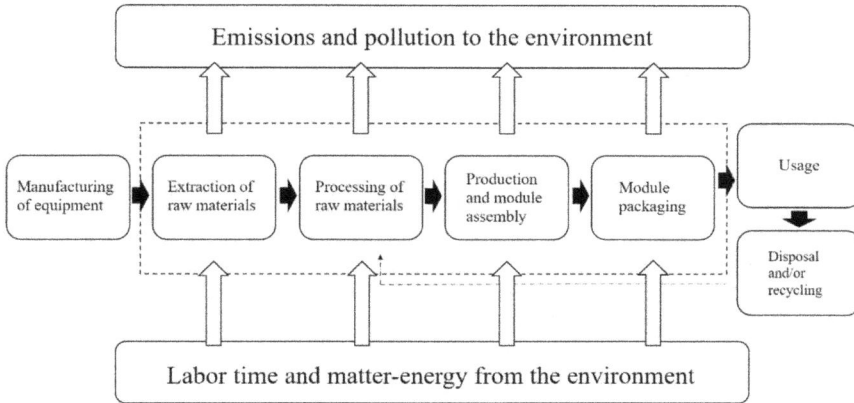

Figure 4.3 System boundary for a Chinese multi-crystalline silicone (m-Si) solar PV module.

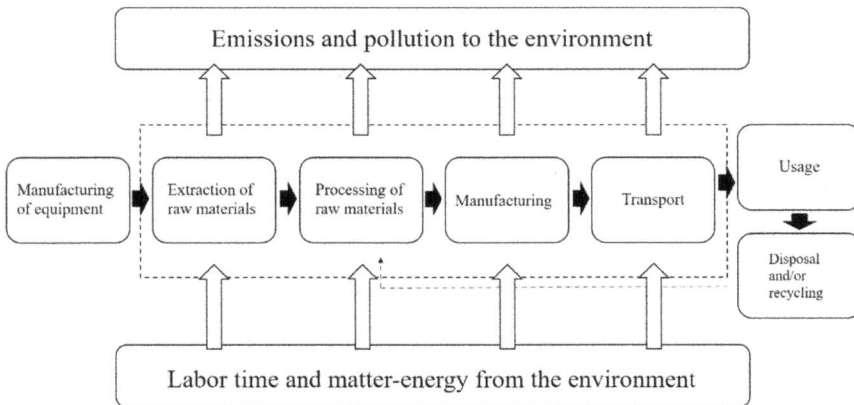

Figure 4.4 System boundary for a German solar photovoltaic manufacturing machine.

was during the last years of consideration, the inventory table is based on LCA analyses published as recently as possible (typically from 2013 onwards). Some LCA analyses show that resource efficiencies in PV manufacturing may not have changed considerably over the last 20 years (Ludin et al., 2018). This may be linked to observations that efficiencies may be offset by indirect energy dissipation and "diseconomies of space" (Bunker and Ciccantell, 2005; Ciccantell and Smith, 2009). I rely on conservative estimations of the resources intensities, i.e., low resource intensities for Chinese solar PV modules and high resource intensities for German solar PV equipment.

Table 4.1 Life-cycle inventory table for a Chinese solar multi-crystalline silicone (m-Si) solar photovoltaic module

Process/input	Energy (MJ)	Land $(m^2)^c$	Labor $(h)^d$	Emissions (kg CO_2-eq.)
Extraction of raw materials	2,552[a]	57	0.35	0.6[b]
Processing of materials	2,071[a]	44	0.87	186.1[b]
Production and assembly	1,382[a]	29	3.26	88.9[b]
Module packaging	1,512[b]	31.5	N/A	167.6[e]
Total	7,517	161.5	4.47	443.2

a   Data from Wong et al. (2016).
b   Data from Dong et al. (2015b).
c   Calculations based on energy values from Wong et al. (2016) and Dong et al. (2015b) converted into land hectares with a coefficient (1.56 W/m^2) calculated from land requirements of China's solar PV industry (Table 5.4). The coefficient includes indirect land requirements of the necessary labor and capital, but excludes the land for carbon sequestration.
d   Data from Llera et al. (2013). Considering a 1,800 h work year in China.
e   Data from Dong et al. (2015b). Includes some emissions associated with assembly of the aluminum frame not previously included.

Table 4.2 Life-cycle inventory table for a German solar photovoltaic manufacturing machine

Process/input	Energy $(MJ)^a$	Land $(m^2)^b$	Labor (h)	Emissions (kg CO_2-eq.)^d
Extraction and processing of raw materials	1,286,500	8,276	N/A	74,617
Manufacturing	61,000	385	N/A	3,538
Transportation	3,225	20	N/A	187
Total	1,350,725	8,681	849[c]	78,342

a   Data from Bochtis et al. (2019).
b   Calculations based on energy values from Bochtis et al. (2019) converted into land hectares with a coefficient (4.93 W/m^2) calculated from land requirements of Germany's solar PV industry (Table 5.6). The coefficient includes indirect land requirements of the necessary labor and capital, but excludes the land for carbon sequestration.
c   Calculated by dividing the annual turnover of Germany's manufacturing and equipment industry with the amount of jobs focused on export in the sector (Kolbe, 2011; Dauth et al. 2017). Considering a 1,350 h work year in Germany.
d   Calculated with a coefficient of Germany's carbon intensity (0.058 kg/MJ) (BP, 2019; Worldometer, 2020).

Table 4.2 summarizes the resource intensity of a German solar photovoltaic manufacturing machine. As previously mentioned, it is notoriously difficult to access LCA data on machinery, perhaps because there is a limited interest in understanding machines as material artifacts (see Chapter 2). The most significant exception to this trend can be found in research on energy flows in agricultural systems. In this field, energy embodiments associated with the manufacture of machinery are considered relevant for the overall energy expenditure of a particular food product or a particular agricultural system. Usually, such measurements are based on energy intensities per kg of machinery. In literature on ecological footprints, it is also possible to find coefficients concerning land embodiments for different industrial sectors (e.g., Hubacek and Giljum, 2003). The life-cycle inventory for the German manufacturing machine draws upon these studies.

The next step is to apply the resource intensities associated with each commodity to aggregate trade volumes between Germany and China during the rise of solar power (see Appendix A). The results show an unequal exchange whereby net transfers of embodied energy, embodied land, embodied labor, and embodied emissions are flowing from China to Germany (Figures 4.5–4.8). In all embodiments calculated, the biophysical exchange implicated in trade with these two commodities shows a notable asymmetry as of 2013. Even prior to this period, a notable (yet smaller) asymmetry was present, which accelerated in 2006. Figure 4.9 provides a closer look at the trends during the years 2002–2010.

These results indicate that trade in the two solar commodities became increasingly unequal between Germany and China between 2002 and 2018. By trading German solar PV manufacturing machinery (secondary industry, heavy

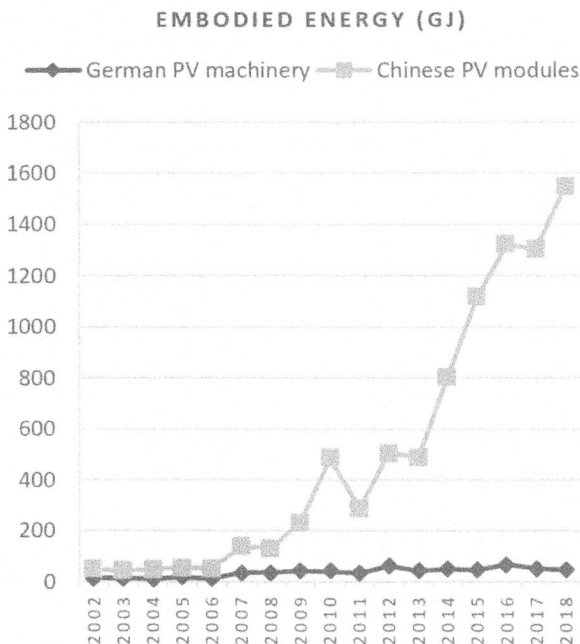

Figure 4.5 Exchange of embodied energy per 10,000 USD.

EMBODIED LAND (HA)

―◆―German PV machinery ―■― Chinese PV modules

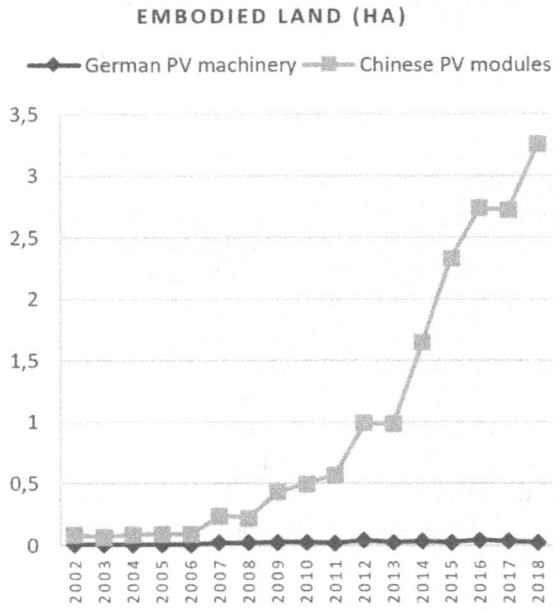

Figure 4.6 Exchange of embodied land per 10,000 USD.

EMBODIED LABOR (H)

―◆―German PV machinery ―■― Chinese PV modules

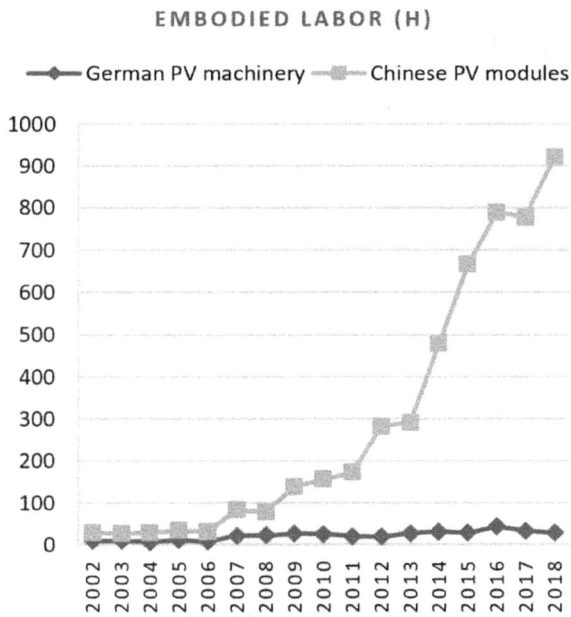

Figure 4.7 Exchange of embodied labor per 10,000 USD.

EMBODIED EMISSIONS (TONNES OF
CO2-EQ.)

—◆—German PV machinery —▩— Chinese PV modules

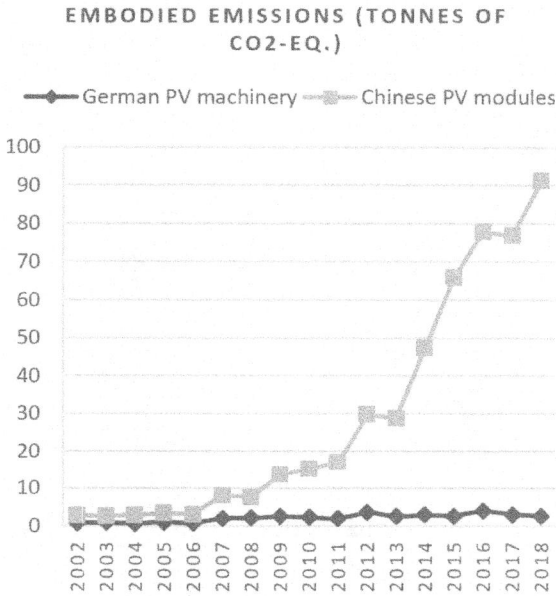

Figure 4.8 Exchange of embodied emissions per 10,000 USD.

EMBODIED LAND (HA)

—◆—German PV machinery —▩— Chinese PV modules

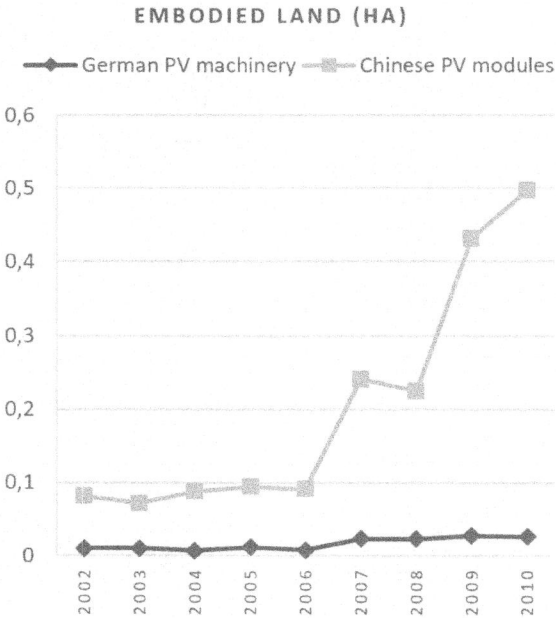

Figure 4.9 Exchange of embodied land per 10,000 USD, 2002–2010.

Table 4.3 Net transfer of embodied resources from China to Germany per 10,000 USD, 2002–2018

Year	Energy (GJ)	Land (ha)	Labor (h)	CO_2-eq. (t)
2002	19	0.1	10	1
2003	14	0.1	8	1
2004	28	0.1	16	2
2005	21	0.1	12	1
2006	27	0.1	15	2
2007	66	0.2	38	4
2008	58	0.2	33	3
2009	146	0.4	85	9
2010	405	0.4	105	11
2011	224	0.5	132	13
2012	382	1	244	23
2013	403	1	238	24
2014	703	1.6	416	41
2015	1,028	2.3	610	61
2016	1,185	2.7	702	70
2017	1,201	2.7	712	71
2018	1,459	3.2	866	86

machinery) for Chinese solar PV modules (secondary industry, light machinery), Germany displaced an increasing amount of the environmental loads of its solar PV development to China. Before the global commercialization of solar PV cells, the exchange implied a modest net transfer of embodied resources per monetary unit. Sixteen years later, the same exchange implicated a more significant net transfer of 1,459 GJ, 3.2 hectares of embodied land, 866 embodied labor hours, and embodied greenhouse gases equivalent to 86 tons of CO_2-eq. from China to Germany per 10,000 USD exchanged (Table 4.3).

The continued fall in the price on solar PV modules in China relative to the comparatively stable price on solar PV manufacturing machine in Germany exemplifies a "deterioration in terms of trade," whereby China needed to export an ever-increasing quantity of solar PV modules to balance the import of solar PV machinery. This is reflected more clearly in how EUE was intensifying during the rise of solar power in 2002–2018 (Table 4.3). To maintain a regular export income flow in the solar PV market, Chinese and foreign manufacturers in China were likely compelled to produce ever-more solar PV modules at an increasing rate of natural resource extraction, pollution, and with lower salaries.[4]

Global asymmetries in solar PV commercialization

The results confirm the previously reported EUE between Germany and China (Tian et al., 2017). Since LCA-based accounting of EUE only measures resource transfers implicated in two commodities, the result here is best understood as an example of a broader pattern of EUE between the countries. Seen from this perspective, the results suggest that Germany's solar PV installation would have implied much higher financial costs and domestic environmental loads if Germany had not engaged in the abovementioned trade with China 2002–2018. But how significant was this ecologically unequal exchange for Germany's prospect to develop its solar PV capacity? This can be tested by calculating the "displacement-adjusted price" and "displacement-based efficiencies" of the solar PV modules.

Let us first turn to the *monetary* costs of the solar PV modules. At the peak of asymmetry in 2018, China exported solar PV modules to Germany at a total exchange value of 463,406,970 USD (Appendix A). The total amount of labor hours embodied was 41,436,627. With Chinese manufacturing wages of 6.2 USD/hour, these labor hours cost 256,907,087 USD (Trading Economics, 2021). With German manufacturing wages of 28 USD/hour, the same amount of labor hours costs 1,160,225,556 USD (Salary Explorer, 2021). The total price on the solar PV modules installed by Germany would have increased to 1,366,725,439 USD (0.74 USD/W) if they were manufactured with German wages. The embodied energy to manufacture the solar PV modules amounted to 69,682,131 GJ (Appendix A). With Chinese electricity prices of 0.08 USD/kWh and German electricity prices of 0.38 USD/kWh (Statista, 2021), the prices on the solar PV modules would increase by 5,806,844,250 USD if they were manufactured in Germany.[5] This means that the solar PV modules that Germany imported from China at an exchange value of 463,406,970 USD would cost 7,173,569,689 USD if they were manufactured with German wages and energy prices. This shows that the price on the solar PV modules would be *at least 15 times higher* if they were produced in Germany rather than China.[6] The displacement-adjusted price is 3.87 USD/W, which is a price equivalent to the market price prior to the boom (see Figure 4.2).

Let us now turn to the *biophysical* efficiencies of Germany's solar PV modules. Considering a capacity of 200 Wp for each solar PV module, the results (Table 4.1) indicate that one solar PV module requires 37.5 MJ, 0.81 m^2, 0.02 labor hours, and 2.2 kg CO_2-eq. per watt installed. In 2018, a given proportion of these resources were appropriated by Germany in trading solar equipment for Chinese solar PV modules. Figures 4.5–4.8 show the environmental load displacement per 10,000 USD. Considering a price of 0.25 USD/W (Figure 4.2), this environmental load displacement amount to 36 MJ, 0.81 m^2, 0.022 labor hours, and 2.22 kg CO_2-eq. per watt installed in 2018. This means that it would have required *40 times more energy, 115 times more land, 32 times more labor, and 34 times more CO_2-eq.* for Germany to install the solar PV modules if they were produced domestically rather than in China.

While the aggregate resource demand would not greatly affect Germany's economy if supplied domestically, the biophysical requirements had a notable influence on the prospects of Germany's large-scale solar PV development by 2018. Let us take the energy efficiency and land efficiency of the solar PV modules as examples to illustrate this. On an average, German solar power generates roughly 3.38 MJ/W per year. Over the course of the 25-year-long lifespan of a solar PV cell, it would generate roughly 85 MJ/W. Considering the biophysical expenditures per watt (Table 4.4), the solar PV module's EROI would be roughly 2.2:1 (85 MJ/W ÷ x).

Table 4.4 Displacement-based efficiency as the difference between biophysical requirements and environmental load displacement

Indicator (per watt)	Energy (MJ)	Land (m²)	Labor (h)	CO₂-eq. (kg)
Biophysical requirement (x)[a]	37.6	0.8075	0.02235	2.22
Environmental load displacement (y)[b]	36.5	0.8005	0.02165	2.15
Displacement-based efficiency (z) = (x − y)	1.1	0.007	0.0007	0.07

a Derived from Table 4.1.
b Derived from Figures 4.5 to 4.8.

This is not high enough to sustain modern industrial societies as suggested by the "law of minimum EROI" (Hall et al., 2009). In contrast, *the displacement-based efficiency suggests that the EROI experienced in Germany was closer to 77:1* (85 MJ/W ÷ z). This EROI is well over the minimum EROI required to sustain modern industrial societies and can largely be attributed to the displacements of energy dissipation to China.

In terms of land, the results show that the solar PV modules required 0.8075 m² per watt (Table 4.4). Dividing this by the 25-year-long lifespan of the solar PV cells (0.8075 ÷ 25) yields 0.0323 m² per watt. Previous studies have estimated that the annual direct land requirement of solar PV utility parks is approximately 5 W/m², i.e., 0.2 m² per watt (Smil, 2015; Capellán-Pérez et al., 2017). Adding these annual direct land requirements (0.0323 + 0.2 m²) yields 0.2323 m² per watt, which translates to a power density of 4.3 W/m². This is a power density that is too low for sustaining industrial societies without significant pressures on domestic food supplies, notable habitat loss, and probably increases in land rent (Smil, 2015). In contrast, by displacing most of the land requirements to China the demand on land experienced in Germany was closer to 0.007 m² per watt, i.e., *a displacement-based power density of 143 W/m²*.

Germany's prospect to install large amounts of solar PV modules was greatly improved by trading solar PV machinery for solar PV modules with China. This effect was not immediate, but gradual, with a notable improvement between 2006 and 2018. The displacement-adjusted price shows that this improvement was highly significant as the falling price on solar PV modules on the world market would have been reset to the market price of 2002 if the modules were manufactured in Germany in 2018. The displacement-based efficiencies also show this by revealing that Germany's solar PV development became a biophysically feasible net energy strategy and required significantly less land per watt capacity due to the environmental load displacements to China.

Implications for contemporary solar PV aspirations

This shows how the rise of the solar power may have been contingent on international price differences facilitating an ecologically unequal exchange between

Germany and China. Without this global asymmetry, the solar PV module prices would probably be much too high and the solar PV module efficiencies much too low to be viable in Germany. This validate the theory of EUE suggesting that a global asymmetric transfer of resources is essential for the concentration and function of solar PV infrastructure in the world's core and should not be understood as distinct from "technological progress" and efficiency improvements.

Quite opposite to what proponents of the Experience curve suggests, this shows that lower relative manufacturing costs was a key necessity for scaling up manufacturing in solar PV modules and not the other way around. This can be explained by the fact that an increase in scale (i.e., "scaling up," or "growth," or even "complexity") can only occur with an absolute increase in matter-energy throughput in highly ordered structures (see Georgescu-Roegen, 1975). Such an increase necessitates lower prices on matter-energy inputs for ventures bound within the cycle of capital accumulation. In the case of the global PV boom, the relative price difference in energy, raw materials, and labor between China and Germany served exactly this purpose, of allowing higher energy-matter throughputs at lower prices. Far from exhibiting unique Ricardian comparative advantages in knowledge on how to scale up, firms in China had the absolute advantage of access to cheap and unregulated use of fossil fuels, lower wages, cheap or unregulated land, and cheap materials necessary to produce larger amounts of solar PV modules at low prices (i.e., scaling up production).

We should recall that relative price differences in resources and wages in different geographical regions is a core mechanism through which ecologically unequal exchange occurs (see Chapter 2). This is reflected in how the annual environmental load displacement increased with the successively lower prices on solar PV modules in China relative to German solar PV machinery. As we see between the years 2011 and 2013, even if export (in kg) increased, the volume of embodiments increased per unit of exchange value (Appendix A). This shows that price differences – not amounts of mass exchanged – are the central determinant for the rates of ecologically unequal exchange (see also the years 2004–2006).

Even if the radical drop in prices in Chinese solar PV modules in 2002–2018 is conventionally regarded as a positive trend, it masks an uneven biophysical relation that was as much a necessary condition for the boom as scientific knowledge, knowledge exchange in manufacturing, or strong government support. For simultaneously as extraction and manufacturing soared in China, the German economy reaped the benefit of the increased import of embodied resources that further contributed to the modernization and complexification of its social metabolism. In this way, ecologically unequal exchange enabled the emergence of a global low-cost solar PV market inseparable from underlying questions concerning environmental sustainability and uneven distribution of work, resources and pollution in the world economy.

It is important to note that the rapid rise of the global solar PV market has not been fast enough to meet installation targets set in the Paris agreement (Halveston and Nahm, 2019). Despite reaping the benefits of Chinese low-cost solar PV modules, Germany has only met 50% of its intended targets of installing

98 GW worth of solar PV capacity by 2030 (Enkhardt, 2019). According to an article in *Nature*, an additional 500 GW of solar PV capacity is expected worldwide already by 2023. The same article suggests that "scaling in production … is something that China … is certainly capable of in the future" (Nature Editorial, 2019: 623). As world leaders race for an increase in PV manufacturing and installation, it is important to understand the underlying biophysical relations that are linked to the continuation of such mega-industrialization. The result uncovered in this chapter may shed theoretical light on some of China's changing trade relations with neighboring its countries.

China has articulated one of the world's most ambitious aspirations for a future industrial economy propelled by renewable energy. The degree to which this aspiration will be fulfilled remains unclear but given its record of accomplishment of manufacturing and installing solar PV modules, it should be taken seriously. In a speech to the UN in October 2016, Liu Zhenya, the former chairman of the State Grid Corporation of China, spelled out the vision for the international organization Global Energy Interconnection Development and Cooperation Organization (GEIDCO) whose mission is to develop the installation of a global renewable electricity grid under the flag of sustainable development. As reported by Fialka (2016) in *Scientific American*, Liu Zhenya laid three phases of the global mega project that would eventually "transmit solar, wind and hydroelectric-generated power from places on Earth where they are abundant to major population centers, where they are often not." First, individual nations will redesign their electricity grids. Second, international efforts will be mobilized to build cross-national grids. Third, power lines and undersea cables will be built to connect cross-national grids across the globe. Liu Zhenya continued to argue that this would create a win-win situation in which sunny regions such as Africa and Central America could export clean energy to major cities that have the biggest need for the energy. In the end, Zhenya claimed, "the world will turn into a peaceful and harmonious global village with sufficient energy, green lands and blue sky." Questions concerning matter-energy requirements, their social distribution, and social-ecological impacts were conspicuously absent.

From what we learn in the case of the boom of the solar PV market, it quickly becomes illusory if social-cultural aspirations – such as Liu Zhenya's – are separated from their associated matter-energy requirements, social distribution, and social-ecological impacts in the world. Here it is important to remember the ecologically uneven relations between China and world-economic peripheries that may come to increase in significance and magnitude with such solar aspirations. Studies have shown that the Chinese PV industry will have to import many of the materials used in solar PV manufacturing from elsewhere in order to further scale up its solar PV manufacturing (e.g., Wang et al., 2019). Simultaneously, Chinese firms are relocating their manufacturing facilities to Southeast Asian countries while attempting to create new export markets in semi-peripheral and peripheral countries (Shen and Power, 2017). In line with this, Yu et al. (2016: 471) suggest that new demand may be explored in less-developed regions that could benefit from increased electrification and green growth. The question, however, is what Chinese firms would demand in turn and what biophysical trade relations this would imply. Taken together, the Chinese social-cultural aspirations for a globally interconnected electricity grid harnessing renewable energy and the biophysical implications of such an endeavor points to the increasing relevance of understanding the distributive problems of advanced solar PV technologies and the anticipated new metabolism.

The vision and the reality of the solar boom

This chapter has shown that the solar PV boom was characterized by an intensification of ecologically unequal exchange whereby Germany could displace an ever-increasing amount of the environmental burdens to China. Without this relation, presented as an option due to relative price differences between the nations, Germany would not have been able to fulfill its goals of installing solar PV capacity without compromising the stability of its domestic economy and the condition of its domestic environment. These results support the hypothesis that a net transfer of resources, flowing from one social group to another, is a *necessary* social relation for realizing conventional aspirations to capture energy directly from the sun through the means of solar PV technology.

The vision of solar PV technology as a democratic or appropriate technology has remained separate discussions on ecologically unequal exchange. The reality of the importance of a world division of labor for solar PV development forces us to reconsider this vision. Large-scale solar PV development is today far from an "appropriate technology." In fact, the results in this chapter are suggestive of Ivan Illich's (1973: 11) conviction that a technology that grows beyond a certain scale tend "to frustrate the end for which they were originally designed." Equally distant is the popular vision of local energy democracies facilitating energy security and greenhouse gas mitigation, for example as argued by Herman Scheer and Naomi Klein. These technological visions and their outlook on technology still hold relevance, but only to the extent that the materiality of the global economy is ignored. As I have argued, however, to omit this aspect is to miss a fundamental part of what technology is (Chapter 2). Even if solar PV modules can be installed to facilitate local democratic communities, there is an additional political dimension to solar PV technologies at the scale of the world economy. This is likely to become an ever-greater concern in need of attention in a world economy where major actors continue to advocate for an ever-faster installation of ever-more solar PV technology.

The ecologically unequal exchange between Germany and China also demonstrates how an instrumental notion of technology, exemplified in German policy, was vital for the emerging global asymmetries. The Renewable Energy Act (EEG), geared toward incentivizing mass installation of PV modules was especially important to this end. It rested upon an understanding of renewable energy technologies, such as solar PV modules, as exempt from so-called "external costs" (BMU, 2000: 16–17). It is clear that fossil energy carriers were understood to have problematic socio-ecological impacts that are not reflected in their price. The

EEG was meant to remedy this situation by "internalizing" the "costs" of fossil fuels, while simultaneously making renewable energy technologies more cost-competitive on the market through subsidies (feed-in-tariffs). This solution, however, never took into consideration that renewable energy technologies also have definite socio-ecological impacts in the world, similar to fossil technologies. In effect, as installation of renewable energy technologies were economically incentivized, the socio-ecological impacts associated with solar PV modules increased. In this sense, the narrow view of solar PV technology (i.e., as exempt from a biophysical past) facilitated the global asymmetries between Germany and China. As theorized in Chapters 2 and 3, this demonstrates how the conventional, fetishized understanding of solar PV technology sustains the global relations of industrial capitalism. These are the very same social relations, we might add, from which the fetishized view of technology and commodities emerged.

Notes

1 This is a reworked version of a paper published by the author in the journal *Ecological Economics*, 199 as an open access article under the CC BY license (Roos, 2022).
2 The photovoltaic effect is a physiochemical effect that generates an electric potential when light hits two (or three) layers of material with polarized electric charges (typically silicone doped with phosphorus or boron).
3 Relations are after all characterized by both separation and connection, as I argued in Chapter 2.
4 The cause of this deterioration in terms of trade cannot be explained by increased "knowledge" or "innovation" in Chinese solar PV manufacturing. For example, the gap in manufacturing costs between Germany and China were so large that even the world's most established solar PV companies – such as German Q-Cells – who had been leading PV manufacturing for decades and had maximized opportunities for low-cost manufacturing within Europe, still could not scale up manufacturing to saturate the growing German demand (Nemet, 2019: 118–123). However, it can potentially be explained by the asymmetries in the functioning of the labor markets and the political economy of environmental regulations in Germany and China, respectively (Pérez-Rincón, 2006)
5 The price of the Chinese electricity was 1,548,491,800 USD (69,682,131 GJ * 0.08 USD/kWh), and the price of the Germany electricity was 7,355,336,050 USD (69,682,131 GJ * 0.38 USD/kWh). The difference between them is 5,806,844,250 USD.
6 We should bear in mind that these figures do not include prices on carbon emissions or expenses associated with the land requirements. If these were calculated, it would likely imply that the prices of the solar PV modules would be even higher if they were manufactured in Germany.

References

Apergis, Iraklis, and Nicholas Apergis. 2019. Silver prices and solar energy production. *Environmental Science and Pollution Research* 26: 8525–8432. https://doi.org/10.1007/s11356-019-04357-1.

Bastani, Aaron. 2019. *Fully automated luxury communism*. London, England and New York, NY: Verso.

Beinhart, Larry. 2018. Why China, and not the US, is the leader in solar power. *Aljazeera*. Published 22-08-2018. https://www.aljazeera.com/opinions/2018/8/22/why-china-and-not-the-us-is-the-leader-in-solar-power/. Accessed 03-03-2020.

Bernreuter, Johannes. 2020. *The polysilicon market outlook 2020: Technology, capacities, supply, demand, prices*. Bernreuter research: Polysilicon market reports.

BMU. 2000. *Act on granting priority to renewable energy sources (renewable energy sources act)*. Berlin, Germany: Federal Ministry for the Environment, Nature Conservation and Nuclear Safety.

Bochtis, Dionysis, Claus Aage Grrøn Sørensen, and Dimitrios Kateris. 2019. Energy inputs and outputs in agricultural operations. In *Operations management in agriculture*, edited by Dionysis Bochtis, Claus Aage Grrøn Sørensen, and Dimitrios Kateris, 187–196. London, England: Academic Press.

Bookchin, Murray. 1964[2009]. Ecology and revolutionary thought. *The Anarchist Library*. Published 27-04-2009. https://theanarchistlibrary.org/library/lewis-herber-murray-bookchin-ecology-and-revolutionary-thought. Accessed 24-09-2020

BP. 2019. *BP statistical review of world energy*. 68th ed. London: BP.

Bunker, Stephen G., and Paul S. Ciccantell. 2005. Matter, space, time, and technology: How local processes drives global systems. *Research in Rural Sociology and Development* 10: 23–44.

Capellán-Pérez, Iñigo, Carlos de Castro, and Iñaki Arto. 2017. Assessing vulnerabilities and limits in the transition to renewable energies: Land requirements under 100% solar energy scenarios. *Renewable and Sustainable Energy Reviews* 77: 760–782. https://doi.org/10.1016/j.rser.2017.03.137.

CAT. 2020a. China. *Climate action tracker*. Published 02-12-2019. https://climateactiontracker.org/countries/china/. Accessed 24-09-2020.

CAT. 2020b. Germany. *Climate action tracker*. 30-07-2020. https://climateactiontracker.org/countries/germany/. Accessed 24-09-2020.

Chan, Royston. 2011. China quells village solar pollution protests. *Reuters*. Published 18-09-2011. https://www.reuters.com/article/us-china-solar-plant-protest/china-quells-village-solar-pollution-protests-idUSTRE78H0FL20110918. Accessed 14-04-2020.

Chen, Geoffrey C. and Charles Lees. 2016. Growing China's renewable sector: A developmental state approach. *New Political Economy* 21(6): 1–13.

Chen, Ying, Yuziang Mao, Maoyong Song, Yongguang Yin, Guangliang Liu, and Yong Cai. 2020. Occurrende and leaching of silver in municipal sewage sludge in China. *Exotoxicology and Environmental Safety* 189: 109929. https://doi.org/10.1016/j.ecoenv.2019.109929.

Ciccantell, Paul S., and David A. Smith. 2009. Rethinking global commodity chains: Integrating extraction, transport, and manufacturing. *International Journal of Comparative Sociology* 50(3–4): 361–384. https://doi.org/10.1177/0020715209105146.

Clark, Brett, and John B. Foster. 2009. Ecological imperialism and the global metabolic rift: Unequal exchange and the guano/nitrates trade. *International Journal of Comparative Sociology* 50(3–4): 311–334. https://doi.org/10.1177/0020715209105144.

Commoner, Barry. 1979. *The politics of energy*. New York, NY: Random House.

Dauth, Wolfgang, Sebastian Findeisen, and Jens Suedekum. 2017. *Trade and manufacturing jobs in Germany*. IZA DP No. 10469. Bonn, Germany: IZA Institute of Labor Economics.

Decker, Kris De. 2015. How sustainable is PV solar power? *Low-tech Magazine*. 25-05-2015. https://www.lowtechmagazine.com/2015/04/how-sustainable-is-pv-solar-power.html. Accessed 14-04-2020.

Dong, Wenjuan, Ye Qi, and Stephen Spratt. 2015a. *The political economy of low-carbon investment: The role of coalitions and alignments of interest in the green transformation in China.* Evidence report no 160: Policy anticipation, response and evaluation.

Dong, Yang, Jingru Liu, Jianzin Yang, and Ning Ding. 2015b. Life-cycle assessment of China's multi-crystalline silicone photovoltaic modules considering international trade. *Journal of Cleaner Production* 94: 35–45. https://doi.org/10.1016/j.jclepro.2015.02.003.

Dorninger, Christian, and Alf Hornborg. 2015. Can EEMRIO analyses establish the occurrence of ecologically unequal exchange? *Ecological Economics* 119: 414–418. https://doi.org/10.1016/j.ecolecon.2015.08.009.

Enkhardt, Sandra. 2019. German government wants 98 GW of solar by 2030. *PV Magazine.* Published 09-10-2019. https://www.pv-magazine.com/2019/10/09/german-government-wants-98-gw-of-solar-by-2030/. Accessed 13-04-2020.

Exter, van Pieter, Sybren Bosch, Branco Schipper, Benjamin Sprecher, and René Kleijn. 2018. *Metal demand for renewable electricity generation in the Netherlands: Navigating a complex supply chain.* Leiden: Metabolic, Copper8, and Universitet Leiden.

Fialka, John. 2016. Why China is dominating the solar industry. *Scientific American.* Published 19-12-2016. https://www.scientificamerican.com/article/why-china-is-dominating-the-solar-industry/. Accessed 03-03-2020.

Fischer-Kowalski, Marina, Elena Rovenskaya, Fridolin Krausmann, Irene Pallua, and John R. Mc Neill. 2019. Energy transitions and social revolutions. *Technological Forecasting and Social Change* 138: 69–77. https://doi.org/10.1016/j.techfore.2018.08.010.

Georgescu-Roegen, Nicholas. 1975. Energy and economic myths. *Southern Economic Journal* 41(3): 347–381. https://doi.org/10.2307/1056148.

Giljum, Stefan, and Nina Eisenmenger. 2004. North-South trade and the distribution of environmental goods and burdens: A biophysical perspective. *Journal of Environment and Development* 13(1): 73–100. https://doi.org/10.1177/1070496503260974.

Gilley, Bruce. 2012. Authoritarian environmentalism and China's response to climate change. *Environmental Politics* 21(2): 287–307. https://doi.org/10.1080/09644016.2012.651904.

Grau, Thilo, Molin Huo, and Karsten Neuhoff. 2011. *Survey of photovoltaic industry and policy in Germany and China.* DIW Discussion Papers No. 1132. Berlin, Germany: Deutsches Institut für Wirtschaftsforschnug.

Haegel, Nancy M., Harry Atwater Jr., Teresa Barnes, Christian Breyer, Anthony Burrell, Yet-Ming Chiang, Stefaan de Wolf, et al. 2019. Terawatt-scale photovoltaics: Transform global energy. *Science* 364(6443): 836–838. https://doi.org/10.1126/science.aaw1845.

Hake, Jürgen-Friedrich, Wolfgang Fischer, Sandra Venghaus, and Christoph Weckenbrock. 2015. The German Energiewende—History and status quo. *Energy* 92: 532–546. https://doi.org/10.1016/j.energy.2015.04.027.

Hall, Charles A. S., Stephen Balogh, and David J. R. Murphy. 2009. What is the minimum EROI that a sustainable society must have? *Energies* 2: 25–47. https://doi.org/10.3390/en20100025.

Helveston, John, and Jonas Nahm. 2019. China's key role in scaling low-carbon energy technologies. *Science* 366(6467): 794–796. https://doi.org/10.1126/science.aaz1014.

Hubacek, Klaus, and Stefan Giljum. 2003. Applying physical input-output analysis to estimate land appropriation (ecological footprints) of international trade activities. *Ecological Economics* 44: 137–151. https://doi.org/10.1016/S0921-8009(02)00257-4.

Illich, Ivan. 1973. *Tools for conviviality.* Glasgow, Scotland: Collins.

Infante-Amate, Juan, and Fridolin Krausmann. 2019. Trade, ecologically unequal exchange and colonial legacy: The case of France and its former colonies (1962–2015). *Ecological Economics* 156: 98–109. https://doi.org/10.1016/j.ecolecon.2018.09.013.

Jäger-Waldau, Arnulf. 2019. Snapshot of photovoltaics—February 2019. *Energies* 12: 769. https://doi.org/10.3390/en12050769.

Jorgenson, Andrew K. 2009. The sociology of unequal exchange in ecological context: A panel study of lower-income countries, 1975–2000. *Sociological Forum* 24(1): 22–46. https://doi.org/10.1111/j.1573-7861.2008.01085.x.

Kavlak, Goksin, James McNerney, and Jessica E. Trancik. 2018. Evaluating the causes of cost reduction in photovoltaic modules. *Energy Policy* 123: 700–710. https://doi.org/10.1016/j.enpol.2018.08.015.

Kolbe, Marko. 2011. *Industrial overview 2011: The machinery and equipment industry in Germany.* Berlin: Germany Trade and Invest, No. 13692.

Li, Linuye, Thomas D. Willett, and Nan Zhang. 2012. The effects of the global financial crisis on China's financial market and macroeconomy. *Economics Research International* 2012(1): 961694. https://doi.org/10.1155/2012/961694.

Llera, E., S. Scarpellini, A. Arandra, and I. Zabalza. 2013. Forecasting job creation from renewable energy deployment through a value-chain approach. *Renewable and Sustainable Energy Reviews* 21: 262–271. https://doi.org/10.1016/j.rser.2012.12.053.

Lovins, Amory B. 1976. Energy strategy: The road not taken? *Foreign affairs.* Published 01-10-1976. https://www.foreignaffairs.com/articles/united-states/1976-10-01/energy-strategy-road-not-taken. Accessed 14-04-2020.

Lovins, Amory B. 1979. *Soft energy paths: Towards a durable peace.* New York, NY: Harper and Row.

Ludin, Norasikin Ahmad, Nur Ifthitah Mustafa, Marlia M. Hanafiah, Mohd Adib Ibrahim, Mohd Asri Mat Teridi, Suhaila Sepeai, Azami Zaharim, and Kamaruzzaman Sopian. 2018. Prospects of life cycle assessment of renewable energy from photovoltaic technologies: A review. *Renewable and Sustainable Energy Reviews* 98: 11–28.

Malm, Andreas. 2013. China as chimney of the world: The fossil capital hypothesis. *Organization Environment* 25(2): 146–177. https://doi.org/10.1177/1086026612449338.

Martinot, Eric. 2010. Renewable power for China: Past, present, and future. *Frontiers of Energy and Power Engineering in China* 4(3): 287–294. https://doi.org/10.1007/s11708-010-0120-z.

Mittlefehldt, Sarah. 2018. From appropriate technology to clearn energy economy: Renewable energy and environmental politics since the 1970s. *Journal of Environmental Studies and Sciences* 8: 212–219. https://doi.org/10.1007/s13412-018-0471-z.

Molina, Manuel González De, and Víctor M. Toledo. 2014. *The social metabolism: A socio-ecological theory of historical change.* Cham, Germany: Springer.

Mulvaney, Dustin. 2019. *Solar power: Innovation, sustainability and environmental justice.* Oakland: University of California Press.

Muteri, Vincenzo, Maurizio Cellura, Domenico Curto, Vinenzo Franzitta, Sonia Longo, Marina Mistretta, and Maria Laura Parisi. 2020. Review on life cycle assessment of solar photovoltaic panels. *Energies* 13: 252. https://doi.org/10.3390/en13010252.

Nahm, Jonas, and Edward S. Steinfeld. 2014. Scale-up nation: China's specialization in innovative manufacturing. *World Development* 54: 288–300. https://doi.org/10.1016/j.worlddev.2013.09.003.

Nature Editorial. 2019. China brings solar home. *Nature Energy* 4(8): 623.

Nemet, Gregory. 2019. *How solar energy became cheap: A model for low-carbon innovation.* London, England and New York, NY: Routledge.

Oulu, Martin. 2015. The unequal exchange of Dutch cheese and Kenyan roses: Introducing and testing an LCA-based methodology for estimating ecologically unequal exchange. *Ecological Economics* 119: 372–383.

Pearce, Joshua M. 2009. Optimizing greenhouse gas mitigation strategies to suppress energy cannibalism. *Second climate change technology conference*, May 12–15, Hamilton, Ontario, Canada.

Peng, Shuijun, Wencheng Zhang, and Chuanwang Sun. 2016. "Environmental load displacement" from the north to the south: A consumption-based perspective with a focus on China. *Ecological Economics* 128: 147–158. https://doi.org/10.1016/j.ecolecon.2016.04.020.

Pérez-Rincón, Mario Alejandro. 2006. Colombian international trade from a physical perspective: Towards an ecological "Prebisch thesis." *Ecological Economics* 59(4): 519–529. https://doi.org/10.1016/j.ecolecon.2005.11.013.

Piano, Samuel L., and Kozo Mayumi. 2017. Toward an integrated assessment of the performance of photovoltaic power stations for electricity generation. *Applied Energy* 186: 167–174. https://doi.org/10.1016/j.apenergy.2016.05.102.

Pomeranz, Kenneth. 2000. *The great divergence: China, Europe, and the making of the modern world economy*. Princeton and Oxford, England: Princeton University Press.

Quitzow, Rainer. 2015. Dynamics of a policy-driven market: The co-evolution of technological innovation systems for solar photovoltaics in China and Germany. *Environmental Innovation and Societal Transitions* 17: 126–148. https://doi.org/10.1016/j.eist.2014.12.002.

REN21. 2016. *Renewables 2016: Global status report*. https://www.ren21.net/wp-content/uploads/2019/05/REN21_GSR2016_FullReport_en_11.pdf.

REN21. 2019. *Renewables 2019: Global status report*. https://www.ren21.net/wp-content/uploads/2019/05/gsr_2019_full_report_en.pdf.

REN21. 2022. *Renewables 2019: Global status report*. https://www.ren21.net/wp-content/uploads/2019/05/GSR2022_Full_Report.pdf.

Reuters. 2017. EU countries oppose duty extension on Chinese solar panels. *Euractiv.* https://www.euractiv.com/section/energy/news/eu-countries-oppose-duty-extension-on-chinese-solar-panels/. Accessed 20-02-2020.

Rodriguez, Miguel, Mario Pansera, and Pablo Cabanelas Lorenzo. 2020. Do indicators have politics? A review of the use of energy and carbon intensity indicators in public debates. *Journal of Cleaner Production* 243: 118902. https://doi.org/10.1016/j.jclepro.2019.118602.

Roos, Andreas. 2022. Global asymmetries in the rise of solar power: An LCA-based account of ecologically unequal exchange between Germany and China 2002–2018. *Ecological Economics* 199: 107484. https://doi.org/10.1016/j.ecolecon.2022.107484.

Roselund, Christian. 2016. Trina solar begins module production in Thailand. *PV Magazine.* Published 29-03-2016. https://www.pv-magazine.com/2016/03/29/trina-solar-begins-module-production-in-thailand_100023919/#ixzz47Nb5oGHs. Accessed 20-02-2020.

Salary Explorer. 2021. Factory and manufacturing average salaries in Germany 2021. http://www.salaryexplorer.com/salary-survey.php?loc=81&loctype=1&job=33&jobtype=1. Accessed 08-03-2021.

Scheer, Hermann. 2005. *The solar economy: Renewable energy for a sustainable global future*. London, England and Sterling, VA: Earthscan.

Scheer, Hermann. 2007. *Energy autonomy: The economic, social and technological case for renewable energy*. London, England and Sterling, VA: Earthscan.

Schumacher, Ernst F. 1993[1973]. *Small is beautiful: A study of economics as if people mattered*. London, England: Vintage.

Shandra, John M., Michael Restivo, and Jamies M. Sommer. 2019. Appetite for destruction? China, ecologically unequal exchange, and forest loss. *Rural Sociology* 85(2): 1–30. https://doi.org/10.1111/ruso.12292.

Shen, Wei. 2017. Who drives China's renewable energy policies? Understanding the role of industrial corporations. *Environmental Development* 21: 87–97. https://doi.org/10.1016/j.envdev.2016.10.006.

Shen, Wei and Marcus Power. 2017. Africa and the export of China's clean energy revolution. *Third World Quarterly* 38(3): 678-697. https://doi.org/10.1080/01436597.2016.1199262.

Sinn, Dann. 2019. Companies look to shift supply chain to Southeast Asia. *Citigroup Inc.* Published 01-06-2019. https://www.citibank.com/commercialbank/insights/asean-supply-chain-changes/en/?linkId=69579292. Accessed 14-04-2020.

Smil, Vaclav. 2015. *Power density: A key to understanding energy sources and uses.* Cambridge, MA: MIT Press.

Smil, Vaclav. 2016. *Energy and civilization: A history.* Cambridge, MA and London, England: The MIT Press.

Stamford, Laurence, and Adisa Azapagic. 2018. Environmental impacts of photovoltaics: The effects of technological improvements and transfer of manufacturing from Europe to China. *Energy Technology* 6: 1146–1160. https://doi.org/10.1002/ente.201800037.

Statista. 2021. Global electricity prices for households in 2020, by select country. *Statista.* Published 03-01-2021. https://www.statista.com/statistics/263492/electricity-prices-in-selected-countries/. Accessed 08-03-2021.

Strumsky, Deborah, José Lobo, and Joseph Tainter. 2010. Complexity and the productivity of innovation. *Systems Research and Behavioral Science* 27(5): 496–509. https://doi.org/10.1002/sres.1057.

Tian, Xu, Rui Wu, Yong Geng, Raimund Bleischwitz, and Yihui Chen. 2017. Environmental and resource footprints between China and EU countries. *Journal of Cleaner Production* 168: 322–330. https://doi.org/10.1016/j.jclepro.2017.09.009.

Trading Economics. 2021. China average yearly wages in manufacturing. *Trading Economics.* Published 09-02-2021. https://tradingeconomics.com/china/wages-in-manufacturing. Accessed 09-02-2021.

Trend Economy. 2020. https://trendeconomy.com/. Accessed 14-04-2020.

UN Comtrade Database. 2021. https://comtrade.un.org/data/. Accessed 03-2021.

UNEP. 2011. *Decoupling natural resource use and environmental impacts from economic growth.* Villars-sous-Yens, Switzerland: UNEP/Earthprint.

UNEP. 2020. *Global trends in renewable energy investment 2020.* Frankfurt am Main: BloombergNEF.

Wang, Peng, Li-Yang Chen, Jian-Ping Ge, Wenjua Cai, and Wei-Qiang Chen. 2019. Incorporating critical material cycles into metal-energy nexys of China's 2050 renewable transition. *Applied Energy* 253: 113612. https://doi.org/10.1016/j.apenergy.2019.113612.

Wang, Zhaohua, and Chao Feng. 2014. The impact and economic cost of environmental regulation on energy utilization in China. *Applied Economics* 46(27): 3362–3376. https://doi.org/10.1080/00036846.2014.929629.

Watt, Jarrod, Naomi Ng, and Finbarr Bermingham. 2019. Behind the tariffs: Solar cells and the exodus of Chinese companies from China. *South China Morning Post.* Published 25-06-2019. https://www.scmp.com/podcasts/article/3015831/behind-tariffs-solar-cells-and-exodus-chinese-companies-china. Accessed 20-02-2020.

WB. 2017. The growing role of minerals and metals for a low carbon future. World Bank Group and EGPS. June 2017. Washington, DC: World Bank Publications.

Wong, Joeson H., Mohammad Royapoor and Chianwen Chan. 2016. Review of life cycle analyses and embodied energy requirements of single-crystalline and multi-crystalline silicon photovoltaic systems. *Renewable and Sustainable Energy Reviews* 58: 608–618. https://doi.org/10.1016/j.rser.2015.12.241.

Worldometer. 2020. German energy. Published 15-03-2020. https://www.worldometers. info/energy/germany-energy/. Accessed 15-03-2020.

WWF. 2020. *Living planet report 2020: Bending the curve of biodiversity loss.* Gland, Switzerland: WWF.

Yang, Hong, Zianjin Huang, and Julian R. Thompson. 2014. Tackle pollution from solar panels. *Nature* 509: 563. https://doi.org/10.1038/509563c.

Yu, Hyun Jin Julie, Nathalie Popiolek, and Patrice Geoffron. 2016. Solar photovoltaic energy policy and globalization: A multiperspective approach with case studies of Germany, Japan, and China. *Progress in Photovoltaics: Research and Applications* 24: 458–476. https://doi.org/10.1002/pip.2560.

Yu, Yang, Kuishuang Feng, and Klas Hubacek. 2014. China's unequal ecological exchange. *Ecological Indicators* 47: 156–163. https://doi.org/10.1016/j.ecolind.2014.01.044.

Yue, Dajun, Fengqi You, and Seth B. Darling. 2014. Domestic and overseas manufacturing scenarios of silicon-based photovoltaics: Life cycle energy and environmental comparative analysis. *Solar Energy* 105: 669–678. https://doi.org/10.1016/j.solener.2014.04.008.

Zhang, Sufang, and Yongxiu He. 2013. Analysis on the development and policy of solar PV power in China. *Renewable and Sustainable Energy Review* 21: 393–401. https://doi.org/10.1016/j.rser.2013.01.002.

5 The inherent politics of global solar technology

Drawing boundaries is a powerful political act. Consider the Enclosure Acts in England (circa 1600–1900) that gave rise to the proletariat through widespread privatization of the commons, or the Valladolid debate in Spain in the mid-16th century that came to justify war against Native Americans on the erroneous belief that they were categorically different from Christian Europeans (so-called "natural slaves"). In both cases, a categorical boundary was drawn and enforced to justify social relations of power, i.e., capitalism and colonialism, respectively. Importantly, there was nothing inherent in the English landscapes that forced the landowners to draw the new property boundaries. Nor were there any inherent biological differences between Americans and Europeans that could justify colonial exploitation and war. The resulting social relations – proletariat/capitalist and slave/colonizer – were not material categories, but socially and historically constructed ones. In the context of drawing boundaries for political purposes, it is interesting to see how cognitive scientists distinguish what they call "artificial" versus "natural" categories:

> By natural object categories we mean categories with lexical entries (e.g., bird) whose instances correspond to entities in the world (e.g., robin, turkey, pigeon). Artificially constructed categories typically involve novel stimuli where the constituent features or properties of examples are familiar, but where the experimenter [or boundary drawer] specifically manipulates the properties of examples and the assignment of examples to categories to create some particular category structure of interest. Neither the examples nor the categories need necessarily correspond to real-world entities (Medin and Heit, 1999: 101–102).

In the examples above, two artificially constructed categories (property and natural slave) were developed for political purposes (capital accumulation and resource exploitation), through the drawing and enforcing of boundaries (enclosure and othering) that were not naturally encouraged in the world. The possibility that humans can cognize categories that have no counterparts in the physical world is at once helpful and politically charged. On the one hand, it allows for deductive learning and hypothesis testing (i.e., establishing categories that are then tested).

DOI: 10.4324/9781003292319-5

On the other hand, in the lack of relations to "real-world contexts," artificial categories can miss or purposefully leave out important aspects of the world, sometimes with far-reaching political consequences (Medin and Heit, 1999: 103).

An illustrative example of the latter is the partition of the African continent during "the scramble for Africa," where European colonial powers during the Berlin congress 1884–1885 drew artificial boundaries to agree over what colonial power owned what territory. This was done with little regard for the real-world context, such as physical features of landscapes (including rivers, forest areas, and mountains) or already established social-political entities or cultural groups (Griffiths, 1986). The awkward straight lines of many African states today is but one reminder of the scant regard in colonial Europe for the African continent and its peoples. Still today, more than a hundred years later, Africa's inherited political geography influences its politics and serves as a catalyst for numerous political-ecological conflicts, including trans-boundary resources disputes (Okumu, 2010) and ethnic wars (Michalopolous and Papaioannou, 2016).

The latter examples demonstrate Alfred Korzybski's (2000[1933]: 58) observation that "a map is not the territory it represents." That is to say, the representation (or categorization) of a thing is not the same as the thing itself. However, the representation (the map) can be truer to the thing (the territory) the more attentive one is to the real-world context. What aspects of reality should be included in the representation, however, is a notoriously political question. Current debates regarding the feasibility of "green growth" illustrates this very clearly. Evaluations of the notion of "green growth" are intimately tied up in questions concerning the categorization of economies, where "decoupling" of economic activity from matter-energy throughput is more likely experienced if the boundary of what constitutes an economy is narrowly drawn at the regional or national level and/or if indirect biophysical costs are omitted from the analysis (Jiborn et al., 2018; Kan et al., 2019). This is a situation in which two sets of categorizations of what constitutes "the economy" typically lead to two sets of conclusions regarding the phenomenon of "decoupling" and the sustainability of "green growth." Whereas the more narrowly defined categorization – of the economy as a national or regional entity – sometimes confirms the presence of relative decoupling, the more widely defined categorization – of the economy as an international phenomenon – tends to disprove it. The logical conclusion, as pointed out in several studies, is that national or regional decoupling occurs, but that it occurs through mechanisms of environmental load displacement, such as ecologically unequal exchange (Jorgenson and Clark, 2012; Isenhour and Feng, 2014). This conclusion can only be reached by considering the economic process as a wider, international category. However, this is methodologically foreclosed in most studies on decoupling.[1]

The question of how to draw boundaries around what constitutes technology (or, technologies) are well known, albeit contested, within philosophy of technology (see Chapter 2). The instrumental definition of technologies, as neutral tools to be employed for different social-political purposes, has been thoroughly criticized. One path-breaking study in this regard, edited by Wiebe Bijker et al. (1989), established that technologies are deeply social both in how they arise from and shape social interactions. The instrumentalist view, from this perspective, is objectivistic and narrow because it captures only a minuscule proportion of what constitutes the vital components of a particular technology. From this critique emerged the systems view of technology, primarily related to the work of Thomas Hughes (1989), who extended the boundary of technological artifacts by recognizing how they were inseparable from a system consisting of organizations and things such as "manufacturing firms, utility companies, and investments banks ... books, articles, and university teaching and research programs" as well as "regulatory law" and "natural resources" (ibid. 51). Crucially, the interaction of each component was understood to "contribute directly, or through other components, to the common system goal" (ibid.). Technological artifacts, in this sense, are component parts emerging from and within a specific systemic telos.[2]

The widely accepted systems view of technology redefines in what sense technologies are considered political. As suggested by Dusek (2006: 36), if we include "advertising, propaganda, government, administration, and all the rest, it is easier to see how the technological system can control the individual, rather than the other way around."[3] The choice of including or excluding certain aspects of technological systems is a political act because it defines who or what is relevant to consider. To understand this, it is worth quoting Hughes (1989: 55) at length:

> The definer or describer of a hierarchical system's choice of level of analysis from physical artifact to world system can be noticeably political. For instance, an electric light and power system can be so defined that externalities or social costs are excluded from the analysis. Textbooks for engineering students often limits technological systems to technical components, thereby leaving the student with the mistaken impression that problems of system growth and management are neatly circumscribed and preclude factors often pejoratively labelled "politics." On the other hand, neoclassical economists dealing with production systems often treat technical factors as exogenous. Some social scientists raise the level of analysis and abstraction so high that it matters not what the technical content of a system might be.

On the one hand, then, technologies can be so narrowly defined that aspects of great importance are omitted from the analysis. On the other hand, technologies can be so broadly defined that they lose sight of any meaningful relation between the system telos and the technologies within it. Winner (1980: 122) warned early that "the corrective," i.e., to draw the technological boundary very wide, could lead to the erroneous conclusion that technical artifacts do not matter at all. Somewhat provocatively, he contends that this conclusion validates what social scientists tend to think of the study of technology in general, "namely, that there is nothing distinctive about the study of technology in the first place" and that they therefore "can return to their standard models of social power ... and have everything they need" (ibid. 122). In response to this, Winner (1980) has offered the to-date most thorough analysis of how political artifacts themselves can be considered political.

Winner's (1980) thoughts on the politics of artifacts can be presented as three ways in which technologies can be political.[4] First, a technological artifact can acquire politics by virtue of being nested in a particular political context that is not necessary for its existence. This is a situation in which tweaking the design, rearranging the political context, or mitigating a particular effect of the technology does not change the device in any significant way. Technologies, in this sense, can nevertheless be political because their existence can be linked to the political aims of the owners or designers (Winner, 1980: 130). In such cases, the technological artifact is both an embodiment of and a tool for the establishment of a particular social relation, but it is not inherently so, since it might exist as a "roughly similar device" also within other social-political contexts. Winner does not name this form of politics, but we could simply call it "acquired" politics on the basis that the particular politics is granted exogenously by the system telos, e.g., from the organizational networks that are in control of the design process (see, e.g., Feenberg, 1991). In this sense, technologies are political in the sense that they can – but do not have to – be employed to further specific social-political relations or agendas (e.g., research, commerce, war, or capital accumulation).

Second, Winner's analysis reveals the existence of so-called "inherently political technologies" that are contingent upon specific social relations of power.[5] Winner (1980: 130) divides these inherently political technologies into two subcategories. For ease of reference, we may call them "strong inherent politics" and "weak inherent politics."

- *Strong inherent politics*: Technologies that require particular social-political relations to exist in the physical world. Railways, nuclear power, and atom bombs, Winner suggests, are technologies that require engineering and military experts operating under hierarchical chains of command that make sure the system is predictable and safe (ibid. 130–132). These technologies require a "social environment to be structured in a particular way in much the same sense that an automobile requires wheels in order to turn" (ibid. 130). The social-political relation, in short, is a necessary condition (or part) of the technological artifact or technological project. Engels, we can recall, argued that industrialism in general, or what he called "the factory system," embodied such an inherent politics (see Chapter 2).
- *Weak inherent politics*: Technologies that are strongly compatible with particular social-political relations. In this political form, a technology does not strictly require a particular social-political context, but it does encourage it. Here, Winner briefly discusses solar energy technology as an example of a technology that many consider to be strongly compatible with a more democratic and egalitarian society, based upon how it encourages decentralized ways of harnessing energy that is at once more "accessible, comprehensible and controllable" (ibid. 130).

Winner himself is somewhat reserved regarding the weak inherent politics of solar energy technology, because this categorization analytically excludes the system's context through which solar technologies are actualized in the world. That is to say, *the perceived democratic potential of solar energy technologies may depend upon how the boundary of the technology is drawn.* In relation to the case of Katrineholm, the appeal to solar PV cells as a democratic and sustainable energy technology may have been contingent on hiding the fact that the cells were imported from Asian manufacturers (we will return to this issue in Chapter 6). In this chapter, I assess the inherent politics of large-scale solar PV projects as it relates to the perceived boundary of what technology is. The aim of this analysis, following Winner, is not to tease out the trade-offs of a particular technology concerning number of jobs, pollution, or potential revenues in a sort of cost-benefit analysis, but to examine the "important consequences for the form and quality of human associations" as a result of the pursuit of solar PV technology, particularly as it relates to the current formation of the new metabolism (Winner, 1980: 131; see Chapter 3).

This chapter discusses the question of whether and in what sense large-scale solar photovoltaic technology projects are inherently political. This will be considered in relation to two biophysical measurements, those of "energy return on energy investment" (EROI) and "power density." First, I attempt to show how methodological disagreements in the study of EROI illustrate how specific ways of drawing the technology boundary is intertwined with different notions of how solar PV technologies are understood as political. Following this, I turn to the "power density" measure and show that the large-scale solar PV projects proposed by four leading solar countries (China, the US, Germany, India, and Italy) are inherently political by virtue of necessitating displacement of land requirements in the world system. In the discussion, I make the claim that the low EROI and low power density of solar PV technologies should not be understood as something that renders a large-scale transition to solar PV technology impossible, but that it raises important questions of justice, ecological sustainability, and social relations of power in the emerging metabolism.

Questioning solar technology boundaries

All ordered structures, including all living things, require an inflow of energy from their environment in order to sustain themselves in the world (see Chapter 2). Among dissipative structures – identified by the ability to reproduce their own structure – some of this energy is necessarily dissipated in the search for energy. The myriad of organisms in nature, all dissipative structures, represent a great variety of strategies by which energy is dissipated in the search for energy. They are nevertheless all subjected to a common rule, namely that they need to acquire more energy than they dissipate during their search for energy. To be sure, many animals and even some humans skilled in the art of fasting can go long periods without energy inputs. But even if many organisms have ways of buffering energy access, e.g., through fat storage, the fact that they need to acquire more energy than they dissipate remains true over their lifetime. Importantly, it is not enough that the organism obtains the exact same amount of energy that

$$EROI = \frac{\text{ENERGY RETURNED}}{\text{ENERGY INVESTED}}$$

Figure 5.1 A simple rendition of how to measure energy return on energy investments.

it dissipates, since the transformation of one energy form to another necessarily implies a loss of useful energy. The digestion of food, for instance, requires that a certain amount of energy be dissipated. In the case of organisms, some energy also needs to be designated to building their own body (e.g., maintaining healthy cells, growing, healing), providing for the continuation of the species (e.g., reproducing, nursing, courting) and enjoyment (e.g., playing, socializing).

The concept of "energy return on energy investment" (EROI) captures this biophysical condition of dissipative structures and presents it as a measurable ratio (Figure 5.1). In its most elementary form, it is understood as the ratio of the energy returned to the energy invested of a particular energy technology (or energy strategy).

The strength of the measure lies partly in its capacity to test some basic biophysical necessities of energy strategies. As I have already mentioned, every energy technology necessitates a higher energy return than the energy invested in building and maintaining that strategy, simply as a way of surviving. This means that the EROI of any energy technology needs to be larger than 1:1 (expressed as "one to one") to be regarded as an energy strategy with the capacity to harness energy from nature. Hall et al. (2009: 29–30) go as far as suggesting that this imperative should be regarded as an "iron 'law' of evolutionary energetics," the so-called "law of minimum EROI." The fact that dissipative structures need to dissipate energy not only for the sake of acquiring energy, means that the EROI of any viable energy strategy needs to be higher than 1:1. There must be some net energy surplus to sustain other necessary practices of the dissipative structure. Hall et al. (2009: 45) have suggested that 3:1 is the bare minimum EROI for sustaining larger human societies. Recent studies (de Castro and Capellán-Pérez, 2020: 4) suggest that an EROI of 10–15:1 is necessary for sustaining advanced industrial societies, with high quality modern healthcare, well developed transportation infrastructure, and more. Lambert et al. (2013), in a study on the relation between EROI and life quality, conclude that a high EROI in a society correlates with few underweight children, high health expenditure per capita, high access to water among rural populations, and high gender equality.

The non-linear character of the EROI measure means that there is a more significant difference between 2:1 and 5:1 than there is between 20:1 and 50:1. The "net energy cliff" illustrates the non-linear relation between EROI, the percentage of energy required to invest in a specific energy strategy, and the percentage of energy returned to society, or the organism (see Murphy and Hall, 2010: 108). An energy technology with an EROI of 2:1, for example, means that 50% of society's energy is dissipated in the procurement of energy. The remaining 50% can be allocated to other ends. In contrast, an energy strategy with an EROI of 10:1

demands only 10% of society's energy for energy procurement, while the remaining 90% can be allocated to other ends. The difference between 10:1 and 50:1 is merely a few percentages. In line with Carnot's theorem, no matter how high the EROI is, a certain percentage of the energy surplus is always necessary for energy procurement.

The measure of EROI is primarily applied to the study of human energy strategies, i.e., energy technologies, where it serves the purpose of assessing the viability of different energy technologies in relation to particular ends. Since the concept rose to prominence, it has triggered two core debates among researchers. These demonstrate the problematic character of the measure. The first of these debates arose from the concern that the production of biofuels, such as ethanol from corn, might require more energy than the final ethanol fuel can deliver to society (Giampietro et al., 1997; Pimentel, 2003; Pimentel and Patzek, 2005). However, the low EROI meant not simply that corn ethanol and other biofuels might not be able to deliver a net energy to society, but also that they would require energy subsidies from other fuels, such as coal and fossil gas, which would contribute to an increase in greenhouse gas emissions. These studies also pointed out the link between a low EROI and the requirement of large amounts of land under intensive industrial cultivation means that biofuels compete with food production and contribute to significant deforestation and soil degradation. A substantial amount of literature and activism then emerged around understanding and combating the environmental impacts of biofuel production, particularly as they were disproportionally felt in the global peripheries (see, e.g., Shiva, 2008; Dauvergne and Neville, 2010; Hermele, 2014).

Contra to this, Farrell et al. (2006) offered an apologetic view, suggesting that the EROI of biofuels might be higher than the break-even point at 1:1 and that they therefore displayed some positive potential in applications for environmental ends (see also Cleveland et al., 2006). This positive potential, however, would only come under future technological improvements, such as "sustainable agriculture and cellulosic ethanol production" (Farrell et al., 2006: 508). Even with the development of new biofuels and continued claims of technological improvements, biofuels remain the least efficient energy technology of industrial societies to date (see Hall et al., 2014; Chiriboga et al., 2020). Meanwhile, environmental and social-political implications have largely been ignored in favor of instrumentalist preoccupations with artifact improvement.

The second debate concerns the EROI of solar PV technologies. In this debate, the major question has been whether the EROI of solar PV technologies is generally high enough to sustain advanced industrial societies. While Prieto and Hall (2013) were early to demonstrate the importance of this question, the kindle of the recent debate was a study by Ferroni and Hopkirk (2016) showing that solar PV systems in regions with moderate solar insolation (such as Switzerland) had an EROI less than 1:1. In these authors' words, solar PV systems can thereby be understood as "non-sustainable energy sink[s] or a non-sustainable net energy loss" (ibid. 343). This was soon met with a comprehensive response from Raugei et al. (2017) pointing out a number of methodological issues and suggesting that the actual EROI of solar PV systems was in fact an order of magnitude higher than Ferroni and Hopkirk's study showed. The response in turn by Ferroni et al. (2017) rebutted these methodological concerns in a manner that allows us to point out some core issues in the debate.

The debate seems at heart to concern the issue of what boundaries are appropriate for an EROI analysis of solar PV technology. The boundary problem is a well-known problem among EROI scholars, and it featured also in the biofuel debate. However, it arguably became more prominent in the solar debate. If we look at the disagreement more closely, we can see that Ferroni et al. (2017: 499) repeatedly indicate that they think Raugei et al. (2017) adopt a narrow understanding of PV technology by "confin[ing] themselves within unrealistic boundaries" for understanding the transition away from fossil energy. Raugei et al. (2017), on the other hand, seem more concerned with making sure that different energy technologies can be subjected to comparison under the same technological boundary. To put it simply, while Ferroni and Hopkirk's (2016) study is concerned with the biophysical reality of transitioning away from fossil energy, Raugei et al. (2017), in their critique, are concerned with establishing a methodological consistency by drawing a narrow, well-defined, technological boundary. Raugei (2019) has elsewhere made the remark that this narrow boundary is drawn for the purpose of providing clear proposals for policymakers. This exemplifies a case in which the particular aim of the study or research program informs what technological boundary is appropriate (for discussion, see Carbajales-Dale et al., 2015; Palmer and Floyd, 2017).

It has long been recognized within the study of EROI that a more widely drawn technological boundary gives a lower EROI ratio and vice versa. Subsequently, numerous types of technological boundaries have been suggested for different aims. These boundaries vary primarily in how to understand the numerator (energy invested). These include, from narrowest to widest, the following technological boundaries (based on Hall et al., 2014; Capellán-Pérez et al., 2019; de Castro and Capellán-Pérez, 2020):

a   *Standard EROI ($EROI_{st}$)*: This boundary represents the narrowest technological boundary. It includes the direct energy dissipated on site and the indirect energy dissipated in the manufacturing of the infrastructure used on site.

b   *Point of use EROI ($EROIp_{ou}$)*: Apart from the energy dissipated directly on site and in the manufacturing of infrastructure used on site, this boundary includes also the energy required for delivering a certain energy carrier to a particular point of use (including refinery, processing, transportation, and maintenance of electric grid).

c   *Extended EROI ($EROI_{ext}$)*: This boundary considers the energy dissipated for getting, delivering and using a given energy carrier. This boundary also extends further back in the production chain by considering the energy embodied in the necessary labor time and capital investment.

This shows how the very same energy technology can be understood in many different ways depending on what technological boundary is chosen. While this

Table 5.1 Energy return on energy investment for solar PV technology under three different technological boundaries

Technological boundary	$EROI_{st}$	$EROI_{pou}$	$EROI_{ext}$
Solar PV technology	7.7–38:1[a]	3.5:1[b]	0.82–4.5:1[c]

a   Raugei et al. (2012), Bhandari et al. (2015), de Castro and Capellán-Pérez (2020).
b   de Castro and Capellán-Pérez (2020).
c   Weißbach et al. (2013), Ferroni and Hopkirk (2016), de Castro and Capellán-Pérez (2020).

fact is taken as evident within the EROI literature, from a philosophical point of view it is an interesting finding. It means that even the physical performances of technological systems can be understood very differently based on how they are demarcated. The different boundaries above thereby represent different epistemologies, but ultimately suggest routes to an alternative ontology of technology. The first point I wish to make in this chapter is that the boundary of solar PV technology may influence not only its calculated efficiency, but also – by extension – the way in which solar PV technology is understood as inherently political. To begin to understand this, we can look at the results of different EROI calculations of solar PV technology under different technological boundaries.

Table 5.1 shows how the expansion of a technological boundary implies a decrease in calculated EROI ratios. This is an expected result. These different boundaries and their results can be understood as corresponding to different demarcations of the technological continuum (see Chapter 2). All the EROI measures take into consideration the energy dissipated for the manufacturing of the infrastructure of solar PV technologies prior to the actualization of the modules. All the boundaries to some degree preclude fetishization. Even the narrowest $EROI_{st}$ measure is a way of understanding the relation between past energy investments and present and future energy returns. In this way, the concept is an acknowledgment that energy is not created from human ingenuity (i.e., a recognition of the first law of thermodynamics). However, each measure does so differently and under different assumptions regarding technology. Since $EROI_{pou}$ is uncommon in the study of solar PV technology, I will focus on the goals, conclusions, and assumptions derived from $EROI_{st}$ and $EROI_{ext}$ respectively.

$EROI_{st}$

Among those favoring the $EROI_{st}$ boundary, the goal, as we have seen, is often explicitly to offer a comparison between solar PV technologies and fossil energy technologies for policymakers. The results, usually concluding that the EROI for solar PV technologies is fairly high, have so far been that solar PV technologies do indeed offer a viable option for replacing fossil energy carriers and for sustaining advanced industrial societies. However, when confronted with the fact that the technological boundary is narrowly drawn, the replies sometimes reveal fetishized understandings of technological progress. For example, in their critique of the EROI$_{ext}$ boundary, Raugei et al. (2017: 378) write:

> It is crucial to recognize that extending the EROI boundaries beyond the inclusion of the physical inputs required for the production and operation of one unit of energy output from the analysed energy system also gradually shifts the goal of the analysis from the (comparative) assessment of its *intrinsic net energy performance* ... to the assessment of the ability of the analysed system to support the entire societal demand for the type of energy carrier it produces (emphasis added).

This is a view in which the technological artifact is of primary importance because it is believed that it can exist and perform tasks apart from its wider systemic context. The phrase "intrinsic net energy performance" reveals the assumption that energy technologies have productive capacities that should be understood as separate from their wider social-ecological conditions. While the necessary past social-ecological conditions are taken into consideration to some degree, they are narrowly considered.

As in the case of biofuels, the view in which the boundary is narrowly drawn serves the ideological function of covering for potential technological inefficiencies with reference to a taken-for-granted improvement of an exogenous system. If a particular technology shows a low EROI value, then this is not taken as a wider cause for concern but seen as an opportunity for further technological progress. For example, Raugei et al. (2012: 580) argue, "technological improvements are expected to continue providing incremental life cycle energy efficiency gains to the existing PV technologies, and even radically more efficient, third-generation devices might become available in the long run." However, the energy necessary for such technological progress is not considered relevant for the EROI analysis, which remains confined to the study of the energy technology's "intrinsic net energy performance" (ibid. 581). Thus, energy consumption associated with such processes as the design and research for improved cell efficiency, the search and extraction of promising raw materials, added transportation routes, additional refining and processing of materials, added links in the commodity chains, etc., are all seen as independent from the technological progress that they are advocating. Meanwhile, we know that the search and extraction of new and higher quantities of materials for the transition to a new metabolism is having very real and politically laden consequences for communities and ecosystems worldwide (see Chapter 1).

Raugei et al. (2017: 382–383) refute the need for including such "associated environmental externalities" in the EROI measure, while simultaneously stressing the importance of keeping up to date regarding technological improvements and data in the case of "rapidly evolving technologies such as PVs." This selective view is symptomatic of "machine fetishism" whereby the technological artifact (e.g., PV module) and improvements upon it are understood as independent from the wider social-ecological relations and conditions that it necessitates. This, in short, is a view that is blind to the wider political-ecological implications of solar PV technologies and their development during the formation of the new metabolism.

Within the overarching goal of providing recommendations for policymakers, studies employing the $EROI_{st}$ boundary thereby primarily serve the political purpose of uncritically legitimizing the large-scale production of solar PV technologies under any social-ecological condition.

$EROI_{ext}$

Among those who favor the $EROI_{ext}$ boundary, the goal is often directed to the question of whether solar PV technologies can be employed to sustain advanced industrial societies (Prieto and Hall, 2013; Pickard, 2014; Ferroni and Hopkirk, 2016; Capellán-Pérez et al., 2019; de Castro and Capellán-Pérez, 2020). Among these studies, comparison between electricity generated from solar PV technologies and fossil energy is of minor importance, because it is already clear that a transition away from fossil energy is both necessary and desirable. Since the last 200 years of rapid industrialization have proven the viability for fossil energy technologies to support the emergence (but not necessarily continuation) of advanced industrial societies, a comparison is not strictly relevant for the aim of these studies. Rather, what is truly pressing is the question whether renewable energy technologies, such as solar PVs, can support global industrialism. The question at stake is the very continuation of high-energy modernity (see Chapter 3). To understand this, the technology boundaries must necessarily be widely drawn to give a more complete picture of the associated energy investments and necessary energy returns.

The methodological difficulties in studying the energy dissipated along the production chains around the globe remains a complex problem, but the results of studies on the $EROI_{ext}$ of solar PV technologies are nevertheless surprisingly similar. Most studies seem to reach the conclusion that the $EROI_{ext}$, ranging from 0.82:1 to 4:1, is too low for maintaining advanced industrial societies (see Table 5.1). Even among the most favorable estimations (4:1), the net energy available to society would amount to no more than roughly 70% of all energy metabolized (see Figure 5.2). Roughly 30% of all global societal efforts (including all jobs, infrastructure, research, etc.) would need to be geared toward the energy sector in such a scenario.

Several conclusions can be drawn from these findings. Some have concluded that solar PV technologies can be a viable option for sustaining industrial societies, but only if they are complemented with fossil energy carriers (Prieto and Hall, 2013; Hall et al., 2014). This, as they often note, defeats the purpose of rapidly installing solar PV technologies in the hope of mitigating catastrophic climate change. In the scenario of a rapid transition relying upon solar PV technology, the low EROI implies that an enormous amount of energy and materials would need to be metabolized during the transition phase.[6] Such an increase in metabolic throughput is antithetical to greenhouse gas emissions reduction targets aiming to remain below a global warming of 2°C, which is why the rapid transition has to be geared toward a degrowth trajectory that aims to reduce aggregate metabolic throughput (Sers and Victor, 2015). This paradoxical situation, wherein the energy transformation is reliant upon an increased matter-energy throughput problematic for reaching the emissions targets, has been called "the energy-emissions trap" and the "transformation paradox" (Sers and Victor, 2015; Heikkurinen, 2019). Capellán-Pérez et al. (2019: 18) similarly conclude that there is a "trade-off between urgent climate mitigation and viability of the [economic] system," essentially showing that the proposed trajectory for "green growth" based on the Environmental Kuznetz Curve hypothesis is detrimental for mitigating climate change. Studies employing the $EROI_{ext}$ boundary thereby serve the political purpose of questioning the sustainability of the large-scale production of solar PV technologies under specific economic imperatives (e.g., growth).

I would like to suggest that both these interpretations have their shortcomings when it comes to the emerging metabolism and its global politics. On the one hand, the claim that technologies harbor an intrinsic productive potential can only be true from a fetishized perspective that purposefully ignores the wider global politics of the emerging solar PV industry. On the other hand, the assertion that solar PV technologies cannot be installed to further advanced industrial societies under the label "green growth" seems to be at odds with the boom in both the manufacturing and installation of solar PV modules over the last two decades (see Chapter 4). The synthesis of these two conclusions, I would argue, is that the low $EROI_{ext}$ of solar PV technologies can partly be circumvented by select social groups if the associated energy investments are displaced elsewhere. The high $EROI_{st}$ of solar PV technologies can be understood as evidence of this possibility. In such a way, the continued use of fossil energy and the continuation of global environmental inequality would make solar PV technologies viable for certain groups at the expense of other groups in the world system.

Therefore, while it may be true that "inequality makes the metabolic system less efficient," a world division of labor[7] may actually serve to increase the EROI of solar PV technologies in certain parts of the world through displacement of material-, labor-, and energy-intensive production processes to other parts of the world (de Castro and Capellán-Pérez, 2020: 6). As shown in the previous chapter, much of the energy dissipated for the German solar boom came from processes outside the national boundary, notably China. This indicates that a country or social group may enjoy higher energetic benefits from the installation of solar PV systems as long as they can outsource the associated energy and land investments to other countries or social groups.

The main point here is that physical inefficiencies are substitutable for global social inequalities in the transition to solar PV technology. This view, crucially, can only be understood from a widely drawn technological boundary that also takes into consideration world trade in resources and world division of labor. This view is in agreement with the generic conclusions drawn among those studying solar PV technology under the $EROI_{ext}$ boundary but adds the question for *whom* the "green growth" trajectory is problematic. If, as some studies suggest (Isenhour, 2016), the policymakers of core regions of the world experience "green growth" from the fetishized perspective of the $EROI_{st}$ boundary, then this can be understood as a cultural impediment to successfully mitigating climate change at a global scale. The transition would, under such circumstances, seem successful for the globally rich (i.e., from their own perspective), while actually depending on increased energy throughput and greenhouse gas emissions displaced to other parts of the world system.

With reference to the low EROI$_{ext}$ of solar PV technology, it remains unclear whether environmental load displacement is not only possible, but also *necessary* for the success of large-scale solar PV technology projects. That is to ask, given their low physical efficiency, do large-scale solar PV technology projects embody a "strong inherent politics" by necessitating environmental load displacement in the world system? With reference to the definition of "strong inherent politics" above, we can rephrase this question in terms of whether a world division of labor is a necessary condition for (or part of) large-scale solar PV projects in the absence of fossil subsidies. To answer this question, we will leave the realm of energy and turn to the calculation of land requirements for the proposed plans for installing solar PV modules in four countries.

Power density and global solar visions

Apart from requiring energy, all organisms and their energy strategies occupy space in the world. The concept of power density, developed by Vaclav Smil (2006, 2015), formalizes this condition in a simple measure designating the horizontal land area needed per watt of energy (expressed as W/m^2) (Figure 5.2). In Smil's own words, power density is a means of "quantifying [the] power that is received and converted (or that is potentially convertible) per unit of land, or water, surface" (2015: 14–15). Similar to EROI, power density is a powerful indicator because of its simplicity and wide applicability. However, it also hides complexities in how to measure the denominator and numerator, which is dependent upon a wide range of contextual factors (Smil, 2015: 23–40). In the case of the denominator (power), this includes factors such as solar insolation of the particular geographical region, PV cell conversion efficiencies, panel direction, risks of direct damage, entropic degradation over time, and more. These are all factors that in situ reduce the theoretical peak power capacity (Wp) of a particular energy infrastructure. For instance, in the case of a modern wind turbine operating under laws of aerodynamics, the efficiency is never higher than $0.59 * Wp$ (Smil, 2015: 19). In the case of the numerator, variations depend upon whether only the surface area of the solar PV modules or the aggregate surface area, including the spacing between the panels or turbines, is considered (de Castro et al., 2013; Smil, 2015: 36–40). However, the land required for the necessary labor and capital should also be included as relevant for the numerator (Hornborg et al., 2019).

$$\text{POWER DENSITY} = \frac{\text{POWER (ENERGY FLOW PER UNIT OF TIME, W)}}{\text{SURFACE AREA (m}^2\text{)}}$$

Figure 5.2 A simple rendition of how to calculate "power density."

Table 5.2 The power density of selected energy technologies

Energy technology	Power density (W/m^2)
Coal, oil, fossil gas	100–24,000
Nuclear	20–4,000
Solar PV power	9–13
Water and wind power	<10
Biofuels	<1

Based on Smil (2006, 2015).

The measure has so far mostly been used in the study of societal energy technologies for understanding the limits to surface areas in the transition away from fossil energy. At root of many assessments over the global implications of transitioning away from fossil energy lie concerns over the low "power density" of renewable energy technologies. Calculations have shown that renewable energy technologies have a relatively low power density in comparison to fossil energy technologies (Table 5.2). Compared to fossil energy technologies such as coal or oil, renewable energy technologies such as solar PVs require substantially more surface areas per unit of energy generated. This applies even to solar PV technologies, which display the highest power density among renewables.

Following this troubling fact, critical geographers have pointed out that the transition away from fossil energy through mass installation of renewable energy technologies may "necessitate new and uneven power relations over land, energy, and territory that will not necessarily point to 'just' transitions to 'sustainability'" (Huber, 2015: 9; McCarthy, 2015; Huber and McCarthy, 2017). Indeed, studies have shown how the import of biofuels and other "green" commodities produced in the world periphery or semi-periphery sometimes represents significant displacements of environmental loads in the world system (Bonds and Downey, 2012; Hermele, 2014). A global political ecology of biofuels is emerging from nations' efforts to transition away from fossil energy while aspiring to maintain an increasing matter-energy throughput.

With solar PV technologies in mind, Huber and McCarthy (2017: 666, emphasis added) contend that

> the geographies of industrial-scale renewable energy production might involve just as many 'extractive peripheries' or 'sacrifice zones' as current geographies of fossil fuel extraction, while their siting and the distribution of the energy produced there, and of its costs and benefits, would be no less *inherently political.*

To transition away from fossil energy, they continue, would imply that access to land become elevated as "the centre of energy struggles," as it once was under the agrarian regime (ibid.). Notably, this understanding of solar PV technologies as inherently political contrasts to the more common view wherein the social-ecological problems associated with solar PV technologies are classified as "unintended consequences" or injustices in the "solar energy commodity chains" that can be engineered away (Andersen, 2013). In this latter view, there is nothing inherently political with the technologies themselves. So far, however, this has not been empirically tested in relation to large-scale solar PV projects.

Power density extended: Principles and how to calculate it

In order to test whether large-scale solar PV projects are inherently political, the necessary surface area to realize a country's solar aspiration can be converted into a percentage of the entire country's surface area, which can then be related to the aspired percentage of solar PV capacity in the country's energy mix. Theoretically, this will reveal the extent to which large-scale solar PV technology projects can be considered as necessarily embodying a "strong inherent politics" by virtue of requiring an amount of surface area that would seriously restrict the country's economy (e.g., food production). This land requirement might also be higher than is feasibly available within a country's borders (e.g., MacKay, 2009).

Crucially, Smil's calculations of power densities only account for the immediate physical potential of a given technological infrastructure and not its total socio-metabolic "footprint" including indirect spatial requirements. In Smil's account, technologies are drawn to coincide with the physical extent of the infrastructure rather than the total sociometabolic system that reproduces it. This corresponds to the technological boundary of the $EROI_{st}$ measure. We can call it "power density standard" (power density$_{st}$). In relation to the technological continuum, the power density$_{st}$ measure considers a given energy technology narrowly as an object in the present. Still, the physical space covered by an infrastructure for harnessing fossil or renewable energy only represents a fraction of the space required to generate the necessary capital to build solar PV infrastructure. Smil estimates that electricity generation from PV solar modules has a power density$_{st}$ of $10–20$ W/m^2 (Smil, 2015). This figure accounts for the space occupied by the surface of the solar modules only. When the aggregate space demanded by solar parks on site are taken into consideration the power density declines to $4–9$ W/m^2 (Smil, 2015: 49–61). We may call this "power density aggregate" (power density$_{ag}$). This measure takes into consideration spacing between the solar PV modules, the infrastructure's "right of way," access roads, service buildings, and space not used on the site. Refining this boundary for solar PV technology, de Castro et al. (2013) examine six newly constructed solar parks in Europe and North America and arrive at an average power density$_{ag}$ of 4.28 W/m^2. The most land-efficient solar park in Spain reaches a power density$_{ag}$ no higher than 5.55 W/m^2 (Table 5.3). For the calculations of the direct land requirements in this chapter, I will assume an average power density$_{ag}$ of 5 W/m^2. Considering the relation between technological boundary and EROI, I expect to find that power density$_{ext}$ will be lower than both power density$_{st}$ and power density$_{ag}$ (see Table 5.3).

Estimating the direct land requirements as well as the indirect land requirements of the necessary labor and capital required in the global commodity chains

Table 5.3 PV technology and its measured "power density standard" and "power density aggregate"

Technological boundary	Power density$_{st}$	Power density$_{ag}$	Power density$_{ext}$
Solar PV technology	10–20	3.7–5.5	unknown

Sources: de Castro et al. (2013), Smil (2015), Capellán-Pérez et al. (2017).

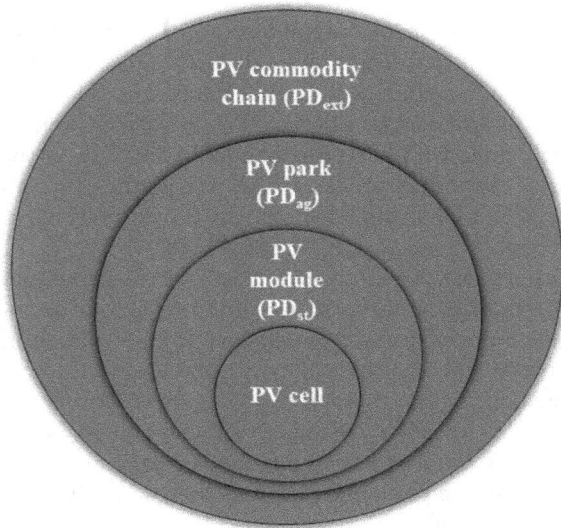

Figure 5.3 The gradually expanding scope of the power density numerator (surface area). Ranging from an instrumental focus on the PV cell to an extended focus considering the global PV commodity chain. PD, power density.

will provide an estimate of what we may call "power density extended" (power density$_{ext}$) (Figure 5.3).[8] This measure, when adjusted to the life span of the solar PV modules, can be used to assess the possible necessity for a country to rely upon a net importation of resources from the global economy in order to construct, install, and maintain a significant PV infrastructure over time.

It is with this in mind that we will turn to the necessary surface area associated with four countries' solar PV aspirations. For generating results that can be discussed in relation to the emerging metabolism, I have chosen five countries based on their significant PV aspirations. These are China, the US, Germany, India, and Italy (see IEA, 2020a). While many countries aspire to expand the share of renewable energy in their national energy mix, governments favoring market-based strategies rarely present detailed plans on how to achieve it, as would be expected. Japan, which is a leader in solar PV installation, has been excluded from the analysis due to the lack of explicit and detailed future energy targets.

The countries analyzed below have either documented their official plans or otherwise outsourced strategy plans to an affiliated organization, which I consult. First, I give a brief description of each country's global energopolitical context, followed by a description of the country's current solar PV aspirations and project plans. Second, I calculate the power density$_{ext}$ in order to relate the land requirements to the available surface area within the country's borders. This allows me, ultimately, to draw some conclusions regarding the "inherent politics" of solar PV technologies, regardless of a specific country's internal social relations of production.

China's aspiration for an "ecological civilization" by 2050

China is by far the most ambitious country when it comes to the installation of solar PV modules. In 2015, China surpassed Germany in total installed PV capacity and since then the expansion of solar PV technology has grown by leaps and bounds. In 2019, China installed more PV capacity than the EU and US combined (IEA, 2020a). Despite recent disruptions in the commodity chain due to the COVID-19 pandemic, China continued to install solar PVs in 2020 and is now expecting a "strong recovery" for solar demand (Bellini, 2020; Shaw, 2020). China is also by far the largest polluter on the planet, currently housing half of the world's coal power capacity and being responsible for emitting almost a third of all greenhouse gases worldwide (Ritchie and Roser, 2017a; CAT, 2020a). China's relentless appetite for coal is only increasing, with an additional 150 GW coal capacity currently under construction domestically and an additional 100 GW financed internationally (Shearer et al., 2019; CAT, 2020a). While the soaring installation of solar PV modules may induce some hope for the environmentally concerned, the latter development is detrimental to any serious effort to mitigate China's – and so the world's – impact on the biosphere (CAT, 2020a). As concluded by Shearer et al. (2019: 4), China's "power generation alone will be more than three times as large as the global limit on coal power use determined by the IPCC to keep warming well below 2°C."

China's proclivity toward fossil energy has put the country under international pressure. This may have been one of the reasons why president Xi Jinping recently pledged that China now plans on becoming carbon neutral before 2060 (Sengupta, 2020). China has previously been participating in international negotiations and pledged to reduce its impacts under the Paris Agreement, even agreeing to a "climate pact" with President Obama to peak its emissions by 2030. While China's targets under the Paris Agreement are reportedly far too modest for maintaining warming below 2°C, Xi Jinping proclaimed that the ambition to peak China's emissions by 2030 will be strengthened (CAT, 2020a). Despite Xi Jinping's recent announcement to the world, the Chinese government has remained silent on how to achieve its ambitions. If Liu Zhenya's speech to the UN in 2016 is any guideline, then the Chinese strategy will operate simultaneously nationally and internationally (see Chapter 4). China is currently active in international mega-industrialization through its "Belt and Road initiative" aiming to connect nations at a world scale through massive expansion of infrastructure, including roads, railways, and sea routes. China's ambitions for a renewable world, as we have seen, also stretch far beyond the borders of China. It remains to be seen, however, whether Zhenya's vision will come to fruition.

While China's international influence is unmistakable, its national targets for phasing out fossil energy through installation of renewable energy technologies is ambitious, but vague. The most concrete formulation of China's ambition can be found in a "roadmap study" by the Energy Research Institute (ERI) of the Chinese National Development and Reform Commission, an executive branch of the Chinese government (c). The study by ERI lays out a "high renewable energy penetration scenario" by 2050 with a stepping stone target for 2030 that aspires to inaugurate an "ecological civilization ... complying with and protecting the nature" (2015: 1). The scenario is highly ambitious, with greenhouse gas emissions peaking before 2025 and the entire economy entering a degrowth trajectory by 2025, effectively proposing to reduce its total energy throughput between 2025 and 2050. The focus of the study is on the electricity sector and solar power and wind power are clearly favored in the proposed scenario. The roadmap suggests that the installation of coal capacity will rise until 2025 and thereafter slowly reduce (perhaps as old plants are decommissioned).

The study sets a goal for the electricity sector for the year 2030, which will be the basis for my calculations. In ERI's (2015: 12) "high renewable energy penetration scenario," 1,048 GW solar capacity will be installed by 2030 and nearly 2,700 GW by 2050. Between 2015 and 2030, a total addition of 999,785 MW, rounded to 1,000 GW, of solar PV capacity is proposed.

If China's trajectory toward an "ecological civilization" progresses in such a way that China succeeds in reaching its 2030 targets, then 25% of its electricity capacity will be made up by solar PV modules by 2030. The success of this target is associated with direct and indirect land requirements amounting to 622,394,950 ha (Table 5.4). To generate 25% of China's total electricity output[9] through this added infrastructure would implicate approximately 63% of China's total surface area.[10] The annual indirect land requirements would amount to roughly 24,095,798 ha during the 25-year-long lifespan of the solar PV modules. Adding this to the direct land requirements (20,000,000 ha) means that an amount of land equivalent to 4.6% of China's total surface area (44,095,798 ha) will need

Table 5.4 Direct and indirect land requirements of ERI's (2015) renewable energy scenario by 2030.[a] Also shows the power density$_{ext}$ calculated

Solar PV	Direct land requirements (ha)	Indirect land requirements, labor (ha)	Indirect land requirements, capital (ha)	Total (ha)
Aggregate	20,000,000	61,443,125	540,951,825	622,394,950
Annual	20,000,000	2,457,725	21,638,073	44,095,798
Power density$_{ext}$	0.16			

a A detailed description of the calculations behind the figures in this table is available in Appendix B.

to be designated to solar PV development every year. Even if the percentage of renewable energy capacity has increased since 2015, solar PV modules covers no more than 1.2% of China's total energy consumption (BP, 2019a). This means that, if carbon sequestration were included within the system's boundaries, it would implicate a surface area more than the entire country to sustain a Chinese economy exclusively propelled by electricity generated from solar PV technology.

The United States' ecological modernization through decarbonization and electrification by 2050

Afflicted by climate denialism fueled by far-right wing politics, the US has been lagging behind in the world's race for renewable energy. The US is the largest producer of fossil fuels in the world, with a national energy mix consisting mainly of oil, coal and fossil gas (Ritchie and Roser, 2022). Despite the Trump administration's absence of effort to change this fossil dependency, the share of primary energy from renewable sources has been increasing since 2010. Renewable energy sources now constitute close to 10% of the US energy supply (Ritchie and Roser, 2022). In 2020, solar power alone stood for 1.35% of its total energy supply (ibid.).

Since the inauguration of the Biden administration, the official vision for the US energy future has dramatically changed. On the day of the inauguration, President Joe Biden signed the executive order to rejoin the Paris climate agreement. By the end of that year, the US submitted its long-term strategy for achieving a net-zero emissions by 2050 (US Department of State, 2021). The administration has then been working on its aspiration to achieve a nation-wide "100% carbon pollution-free electricity by 2035" (The White House, 2022a). These commitments are part of a wider societal vision for ecological modernization, including massive efforts to increase the supply of renewable energy, efforts to increase the production of electric vehicles, efforts to establish domestic supply chains of batteries, all while promising an increase of "clean-jobs" providing the population with the means to uphold its high energy level of consumption.

The current administration's international relations is colored by its commitment to renewable energy through ecological modernization. The administration is working on a long-term goal for clean energy independence – primarily from China and Russia. This includes establishing economically competitive domestic supply chains tripling US domestic production of solar technology (The White House, 2022b) and banning cheap solar imports from the Chinese Xinjiang province (Wagman, 2021). Simultaneously the Biden administration is lifting the tariffs on solar imports from Chinese manufacturers in Cambodia, Malaysia, Thailand and Vietnam, which were enforced until suspicions of waste dumping could be investigated. The reason for lifting the tariffs, ironically, is that cheap solar imports from these countries are necessary for the drastic increase in US domestic solar manufacturing promised by the administration (Sylvia, 2022).

Despite ambitious efforts on behalf of the current US government to decarbonize its economy, it remains unclear whether these efforts are sufficient or

Table 5.5 Direct and indirect land requirements and power density$_{ext}$ of DOE's (2021) "Decarb+E" scenario by 2050

Solar PV	Direct land requirements (ha)	Indirect land requirements, labor (ha)	Indirect land requirements, capital (ha)	Total (ha)
Aggregate	32,000,000	141,400,000	402,675,000	576,075,000
Annual	32,000,000	5,656,000	16,107,000	53,763,000
Power density$_{ext}$	0.28			

realistic. The Climate Action Tracker labels the current US climate targets as "insufficient," as warming would reach 2°C–3°C, if all countries were to follow the US approach (CAT, 2020b). While the US policy is now seeking an increase in its renewable energy capacity, its commitment to reduce its fossil fuel dependency remains too modest and highly reliant upon dubious schemes for carbon removal. Political pressure (from democrats and republicans alike) to ramp up its domestic production of oil following the Russian invasion of Ukraine also demonstrates how essential fossil fuels remain to the US government as a political actor in the world economy. Further uncertainty stems from the country's stark political division on social-ecological sustainability and justice, meaning that current energy policies could be unrolled fast with a shift in government (Bang, 2021). There is also opposition against ecological modernization and its environmental injustices from dissenting voices within the country, including grassroots organizations and indigenous groups, which contribute to shaking the government's high-energy aspiration of a fossil-free nation (Susskind et al., 2022).

The US Department of Energy (DOE, 2021) released a comprehensive study on three different future energy scenarios including a special appraisal of solar power. The scenario "Decarbonization with Electrification (Decarb+E)" is aligned with current government's aspiration to achieve net-zero emissions, while also accounting for the expected increase in electricity generation and demand. With an estimated increase of 50 GW solar capacity annually, the scenario assumes 1,600 GW additional capacity by 2050.

If the assumptions behind these calculations are accurate, then the US's aspiration to install 1,600 GW by 2035 will implicate roughly 576 million ha (Table 5.5). This added capacity, which will make up 45% of the US's total electricity capacity in the Decarb+E scenario, will implicate a surface area equivalent to 58% of the US total surface area.[11] Over the 25-year-long lifespan of the PV modules, the indirect land requirements amount to 21.8 million ha per year. These annual requirements (including the direct land) amount to 53.76 million ha, which is equivalent to roughly 5.5% of US's total surface area. In the Decarb+E scenario, electrification of the US economy would amount to a generous 20-37% of its primary energy share (see "High electrification" in Murphy et al., 2021). The 45% of electricity generated through the added solar PV capacity represents 9-16.5% of the country's expected energy demand in 2050.

Germany's coal phase-out and renewable electricity plan

Germany has long been recognized as a global leader in solar power, only recently surpassed by China in installed solar PV capacity. Similar to China, however, Germany is still heavily reliant upon a range of fossil energy sources, including coal, oil, and fossil gas. The country's increasing fossil dependency has partly been blamed on the phase-out of nuclear power under the *Energiewende* with the aim to transition to a clean and low-carbon energy mix by 2050 (Bruninx et al., 2013; cf. Kunze and Lehmann, 2019). Despite Germany's high success in installing renewable energy technologies, the country relies heavily on energy imports. As much as 64% of Germany's total energy consumption is imported. The country's use of oil and fossil gas are both heavily reliant (at least 97%) upon import, primarily from Russia (Gazprom) and Norway (BMWI, 2020). Oil and fossil gas, moreover, constitute the two largest primary energy carriers for the German economy, making the country highly dependent on the international market for its energy throughput.

In contrast, the largest domestic energy source consists of coal (both lignite and bituminous). Historically, coal was a key energy source for German industrialization and development. Today, while Germany's ambition is to phase out coal, new coal-fired power stations are still being built (Meza, 2020) and the persistent proclivity toward coal for electricity generation is being met by widespread grassroots mobilization and protests.

In efforts to mitigate climate change, ensure energy security, and stave off unrest, the government approved the Climate Action Plan 2050 (*Klimaschutzplan 2050*) in 2016. The plan aims to make the country carbon neutral by 2050, while increasing the share of renewable energy in the energy mix (BMUB, 2016). After the UN summit (COP25) in 2019, Greta Thunberg and 15 other young activists filed a lawsuit against the five biggest polluters, including Germany, for breaching the Convention on the Rights of the Child on the basis that these countries were promoting fossil energy and thereby compromising the lives of current and future children. Perhaps as a response to these accusations, Germany adopted a climate policy package the same year, which included a plan for reaching the interim targets for 2030, lifting the 52 GW PV cap and confirming a Climate Action Law. The most exciting of these developments is arguably the law that aims to phase out coal by 2038, starting with the shutdown of four coal power plants in Rhineland already by the 31st of December, 2020 (Wettengel, 2020).

Until recently, it was believed that Germany would miss its climate targets set for 2020 to reduce greenhouse gases to 40% of its 1999 levels (Enkhardt, 2019a). The Climate Action Tracker, however, projects that Germany will now likely reach its targets as a direct consequence of the reduced industrial production and energy demand during the pandemic (CAT, 2020c). Nevertheless, the efforts of the German government are not deemed compatible with a 1.5°C pathway, largely because the proposed actions come too late, with too many compromises and with too vague guidelines for the project plans (ibid.).

The most concrete scenario projections for Germany's 2050 plan have been developed by the climate think-tank Agore Energiewende, which provides three

Table 5.6 Direct and indirect land requirements in Germany's target scenario "KA65."[a] Also shows the associated power density$_{ext}$

Solar PV	Direct land requirements (ha)	Indirect land requirements, labor (ha)	Indirect land requirements, capital (ha)	Total (ha)
Aggregate	1,000,000	898,956	11,727,725	13,717,681
Annual	1,000,000	35,958	469,109	1,505,067
Power density$_{ext}$	0.36			

a A detailed description of the calculations behind the figures in this table is available in Appendix B.

possible scenarios for Germany's energy mix in 2030 depending on policy decisions (AER, 2018). Out of these, the scenario "coal phase-out + 65% renewable energy" is the one most aligned with the current government's position and promises. In this scenario, an additional 50 GW PV capacity will be installed by 2030, a target explicitly mentioned by the German government (Enkhardt, 2019b). This means, in essence, that solar PV capacity will double over a period of 13 years. This target scenario will serve as the basis for my calculations.

As we have seen, Germany's target is to provide 35% of its electricity from solar PV modules by 2030. Out of this percentage, the added 50 GW would contribute to roughly 17% of Germany's total capacity for electricity generation (half of the total installed capacity in 2030). We can now see that this ambition to install an additional 50 GW of solar PV capacity would require access to 13,717,681 ha, which is equivalent to a power density$_{ext}$ of 0.36 (Table 5.6).

Given that the total surface area of Germany is 357,386 km^2, Germany's ambition to generate 17% of the country's electricity through solar PV capacity would implicate an amount of land equivalent to 38% of its total surface area. Over the course of the 25-year-long lifespan of the solar PV modules, the annual indirect land requirements are 505,067 ha (1/25 of 12,626,681 ha[12]). However, not only the indirect land requirements, but also the direct land requirements will be required each year. The total annual land requirement, 1,505,067 ha, amounts to 4.2% of Germany's total surface area. If we take into consideration that renewable energy supplied 14.8% of Germany's primary energy in 2019, 1.6% of which was generated from solar PVs, this means that it took 4.2% of Germany's land surface to supply 1.6% of its energy in 2019 (AGEB, 2020). Thus, in the case of a scenario in which Germany's entire energy supply comes from renewable energy technologies, such as solar PVs, it would likely require a surface area greater than that of the entire country.

India's path toward energy independency through 2030

The "rise of India" can largely be attributed to the country's massive burning of domestic coal and imported oil (Chikkatur and Sagar, 2009; Chacko, 2015). Today, India is the world's third largest coal producer (after China and the US) and second largest coal consumer (after China), consuming a massive 5,172 TWh in 2019 (Ritchie and Roser, 2017b). The Indian economy is also importantly geared toward burning imported oil from Iraq, Saudi Arabia, and Iran. This is linked to the changing nature of the Indian state in the beginning of the 2000s, which reframed the country's energy situation in terms of "energy security" for economic growth, increased welfare, and international competitiveness (Chacko, 2015). Access to oil and coal were two key energy carriers that became central to achieve this aspiration. Given its very modest oil reserves, India's increased economic growth over the last years has been contingent upon multilateral co-operation in the international arena, including an active involvement in curbing piracy on the Somalian coast in order to secure reliable supplies of oil (Chacko, 2015). However, relying on oil imports and diminishing coal reserves bodes ill for long-term energy security and ecological sustainability (Chikkatur and Sagar, 2009; CAT, 2020d).

In a nutshell, the current focus of the Indian government is to simultaneously expand its coal and renewable energy capacity (Spencer et al., 2020). This may seem like a contradictory pathway, but reflects the government's prioritization of energy security and economic growth in an internationally competitive context. On the one hand, India can be understood as an international leader and role model for the energy transformation under a government that has pledged 500 GW of renewable capacity by 2030, leading to more than 60% renewable energy in its electricity mix. On the other hand, the Indian government's increasing dependence upon coal can be understood as not consistent with the Paris Agreement and possibly detrimental for tackling climate change (Shearer et al., 2017; CAT, 2020d). Regardless of how we interpret this situation, India is set on continuing its economic development through green technology and growth to saturate the rising demand for electricity in its urban centers.

These developments will likely alter the Indian energy landscape as well as international energy politics. As stated by a recent commentator, "if India can convert even a tiny portion of the 150,000 gigawatts of natural solar radiation that it's bombarded with per year into cost-effective electricity, it will not only transform the energy and manufacturing landscape of the world's largest democracy but also dramatically alter the geopolitical equation of Asia" (Sinha, 2020). The hope, as stated in the article, is that India will manage to both manufacture, produce, and consume renewable energy within its borders independently from Chinese manufacturers and interests. This aspiration came partly as a response to the crippling supply chains of Chinese solar PV modules in the wake of the COVID-19 pandemic. For this end, the Modi government is currently making ready land for over 40 solar parks throughout India. The question, however, is whether India's solar ambitions can be implemented in democratic ways. Already, studies have revealed how the development of solar energy mega-projects in India dispossesses vulnerable communities for the benefit of capital coalitions through the enclosure of commons and extra-legal land politics (Yenneti et al., 2016; Stock and Birkenholtz, 2019a, b). Apart from this, it is also unclear to what degree the large-scale development of solar PV technology can fit into a nationally confined energy system not reliant upon international markets.

As in the case of Germany, the Indian government does not provide target scenarios of their own. However, statistics over the government's ambitions can be found in a pathway study provided by the New Delhi-based Energy and Resources Institute (Spencer et al., 2020). In their own words, the point of the study is to "outline what is required to take the share of VRE [solar power and wind power] to levels greater than 30% of generation by 2030" (ibid. 1). The model provides two scenarios, the "Baseline Capacity Scenario" with 36% of electricity generation from renewable energy technologies and the "High Renewable Energy Scenario" with 64% from renewable sources. The latter scenario is aligned with the Modi government's target for 500 GW renewable capacity by 2030. In this scenario, the total amount of renewable generation capacity amounts to 505 GW (adding hydro, wind, solar, and biomass/waste). In analyzing the requirements of these targets, the study reveals that India's future energy mix will likely require a substantial amount of battery storage. Since I do not estimate the land requirements of battery storage in the other cases, I will intentionally leave out the labor and raw materials necessary for this infrastructure and focus only on the direct and indirect land requirements associated with the additional 184.3 GW PV capacity modeled in the "High Renewable Energy Scenario."

If these calculations are accurate, then India's aspiration to install an additional 184.3 GW solar PV capacity will have a total land requirement of 159,882,954 ha (Table 5.7). This added capacity, which will make up 23% of India's total electricity generating capacity, will implicate a surface area equivalent to 49% of India's total surface area.[13] Over the 25-year-long lifespan of the PV modules, the indirect land requirements amount to 6,247,878 ha per year (1/25 of 156,196,954 ha indirect land requirements). Thus, the annual requirements (including the direct land) amounts to 9,933,878 ha, which is equivalent to 3% of India's total surface area. Today, the latest data indicate that solar power in India – including PV and concentrated solar power (CSP) – contributes to 0.9% of the country's total energy consumption (BP, 2019b).

Table 5.7 Direct and indirect land requirements for India's 2030 target.[a] Also shows the power density$_{ext}$ calculated

Solar PV	Direct land requirements (ha)	Indirect land requirements, labor (ha)	Indirect land requirements, capital (ha)	Total (ha)
Aggregate	3,686,000	7,283,904	148,913,050	159,882,954
Annual	3,686,000	291,356	5,956,522	9,933,878
Power density$_{ext}$ (W/m^2)	0.12			

a   A detailed description of the calculations behind the figures in this table is available in Appendix B.

Italy's solar development and commitment to the Belt and Road initiative

Italy has a long history of being dependent upon imported fossil energy. Still today, the country's primary energy supply is made up of over 83.7% fossil energy, primarily in the form of crude oil imported from Azerbaijan, Iraq, Russia, and other countries. In recent years, however, Italy has increasingly turned to renewable energy for satisfying its energy demand, even if the proportion of renewable energy in its primary energy supply remains modest. For a time, during the solar PV boom, Italy was the world's leading nation in solar power. Today, within the EU, it is second only to Germany with an installed capacity of roughly 20,000 MW. Nevertheless, with a large share of fossil energy, Italy's efforts to mitigate climate change rank as "insufficient" alongside the other EU member states (CAT, 2020e).

In the beginning of the 2010s, the two Italian energy corporations Enel Green Power and Terna were part of the Germany-initiated DESERTEC project that was intended to provide electricity to Europe via high-voltage direct current transmission generated in the Saharan desert. The vision-trigger for the massive project was a thesis claiming that only a fraction of the land of the highly insolated Saharan desert could supply the entire world's energy need (May, 2005). The energy need of all the countries in Europe could be saturated from an even smaller surface area. At the time of the proposal, DESERTEC was seen as a promising solution to environmental problems, energy security and peak oil. As one commentator put it, as "an apolitical techno-fix, it promises to overcome these problems without fundamental change, basically maintaining the status quo and the contradictions of the global system that led to these crises in the first place" (Hamouchene, 2020). The non-political essence of the proposals eventually led to splits within the project, separating those who saw it as a way to satisfy European demand and those seeing the project as the beginning of a truly international and democratic power grid. The basis of the failure of the wider project coalition, however, was the marked drop in prices of renewable energy technologies during the time of the solar PV boom (ibid.). Still today, however, some of the DESERTEC partners are in continued collaboration with the aim to generate solar power (mostly through CSP) from the Saharan desert.

With the decline of the DESERTEC project and increased collaboration with Asian markets, Italy is now in close collaboration with Chinese developers working on the massive Belt and Road initiative. In 2019, the Chinese corporation Jetion Solar signed a deal with the Italian state-owned fossil corporation Eni to install 1 GW solar PV power in the country (Hall, 2019). This deal reportedly came as a reward for Italy's agreement to be part of the initiative that will encompass solar PV projects spanning from South East Asia to Europe. In turn, Italy is starting to establish manufacturing facilities in China (Hutchins, 2019). In a stroke of historical repetition, Italy is once again connected to Eastern Asia via the historical route of the "Silk Road." Ironically, then, while the Italian government reportedly sees the European Green New Deal as "an opportunity for

Table 5.8 Direct and indirect land requirements for Italy's 2030 target.[a] Also shows the power density$_{ext}$ calculated

Solar PV	Direct land requirements (ha)	Indirect land requirements, labor (ha)	Indirect land requirements, capital (ha)	Total (ha)
Aggregate	628,760	2,147,627	6,730,426	7,361,959
Annual	628,760	85,905	269,217	983,882
Power density$_{ext}$ (W/m^2)	0.43			

a   A detailed description of the calculations behind the figures in this table is available in Appendix B.

Europe to become a key geopolitical actor," such an opportunity might become contingent upon a global political economy with core actors in Eastern Asia (Coratella, 2020).

In the midst of these developments, three ministries of the Italian government presented the joint Integrated National Energy and Climate Plan (INECP) that targets to supply 55% of the country's electricity generation from renewable energy sources in 2030. To achieve this, Italy plans to develop an additional 31,438 MW PV capacity. This figure is the basis for the calculations below.

In total, the direct and indirect land requirements of Italy's ambition to install 31,438 MW PV capacity amounts to 7,361,959 ha. These figures yield a power density$_{ext}$ of 0.43 (Table 5.8). This added capacity, which represents 18% of Italy's electricity capacity in 2030, would necessitate an amount of land that is equivalent to 24.4% of Italy's total surface area.[14] The indirect land requirements in the necessary labor and capital, however, are dispersed over time. Over the 25-year-long lifespan of the solar PV modules, these land requirements (8,878,053 ha) are equivalent to 355,122 ha per year. If we also consider the direct land requirements as annual, the total annual land requirements (983,882 ha) amount to roughly 2.5% of Italy's total surface area. Today, electricity generation represents 21% of Italy's total energy consumption (IEA, 2020b). To the degree that this remains the same in 2030 (despite INECP's ambitious targets), the 18% of electricity generated through the added solar PV capacity represents 3.8% of Italy's final energy consumption and the associated power density$_{ext}$.

The political ecology of the technological boundary

I will now discuss some of the most central implications of these results. First, the calculations suggest that the power density$_{ext}$ of large-scale solar PV projects is likely somewhere between 0.12 and 0.43 W/m^2. This is an order of magnitude lower than the values reached with the power density$_{ag}$ boundary (Table 5.9). This means that solar PV technology, when understood from a broad systems perspective, requires a substantially higher amount of land per watt capacity. Since no equivalent to the "net energy cliff" has been developed for the power density measure, it is difficult to determine exactly at what value a particular technology

Table 5.9 Solar PV technology and its measured "power density standard," "power density aggregate," and "power density extended"

Technological boundary	Power density$_{st}$	Power density$_{ag}$	Power density$_{ext}$
Solar PV technology	10–20	3.7–5.5	0.12–0.43

Sources: de Castro et al. (2013), Smil (2015), Capellán-Pérez et al. (2017), and the author's calculations.

can be considered to require a land subsidy. In this chapter, I have taken the relation between a country's total surface area and the surface area necessary to provide that country's energy as relevant for understanding the inherent global politics of solar PV technology. The inherent politics of solar PV technology is found in the relation between the social aspirations and the biophysical conditions to fulfill them, i.e., in the mismatch between the socially motivated ends and the biophysical means or limits.

So, do the inherent politics of solar PV technology depend upon how the technological boundary is drawn? Table 5.10 shows the results above in relation to the same calculations under the power density$_{st}$ boundary. Here we can see that the differences between the two boundaries are significant. If we look at electricity generation, the solar aspirations of all countries seem favorable from the perspective of the power density$_{st}$ boundary; to generate between 17% and 25% of the countries' electricity would require only 0.3%–1.4% of their respective total surface area. In contrast, from the perspective of the power density$_{ext}$ boundary, all the aspirations seem more problematic; to generate 17%–25% of a country's electricity supply would necessitate 2.5%–4.6% of a country's total surface area. This means that it could implicate up to 25.6% of an entire country's surface area to generate 100% of the electricity from PV technology. If we recognize that

Table 5.10 Land requirement related to solar PV energy generated from the viewpoint of two technological boundaries

Country	Power density$_{st}$	Power density$_{ext}$
Energy generated by PV (%)	Annual land requirement (%)[a,b]	Annual land requirement (%)[a]
China: 25% of electricity, or, 1.2% of total energy supply	0.5–1	4.6
US: 45% of electricity, or, 9-16.5% of total energy supply	0.81–1.62	5.5
Germany: 17% of electricity, or, 1.6% of total energy supply	0.7–1.4	4.2
India: 23% of electricity, or, 0.9% of total energy supply	0.3–0.6	3
Italy: 18% of electricity, or, 3.8% of total energy supply	0.4–0.8	2.5

a Percentage of the country's total surface area.
b Calculated by considering a power density of 10–20 W/m^2.

electricity only constitutes a small portion of a country's total energy supply, then we can see that there is a significant difference between the boundaries. In all cases but Italy, the percentage of the countries' surface area that is needed for the PV electricity is *higher* than the percentage of PV electricity in the total energy mix. This means that it would require more surface area than the geographical territory of the entire country if 100% of the total energy supply is generated from solar PV technology. Notably, the power density$_{st}$ boundary cannot detect this possibility.

The results confirm that conventional solar aspirations may require more land than the accessible surface area if 100% of the total energy supply is generated through solar PV technology (for a similar conclusion, see MacKay, 2009). At the very least, the results demonstrate that this will be practically difficult without a net import of embodied labor, embodied land, or raw material. This is so even if the indirect land requirements are dispersed over the life span of the solar PV modules. Even if the US and Italy could theoretically keep their solar energy production within their borders, the question is how the very high land requirements would compete with other claims on land and the natural habitat of local species.

The calculations suggest that large-scale solar PV technology projects are probably inherently political by virtue of necessitating or encouraging ecological appropriation in the world system. This means that the large-scale solar PV ambitions of China, the US, Germany, India, and Italy are probably unfeasible if they are not coupled to either fossil subsidies, an ecological appropriation from other parts of the world economy, or both. Since the direct land requirements increase with additional installations of solar PV capacity, this inherent politics of solar PV technology will probably not be mitigated. While some countries, such as Japan, are now installing solar PV modules on water surfaces, this leads to a dynamic whereby increasing installations within the border of a country puts further pressure on that country, thereby encouraging extraction, manufacturing, and labor displacements. Technological progress in cell efficiency, as far as it is dependent upon an increasingly complex commodity chain, is not by default a solution because it is associated with increased biophysical expenses (e.g., new material components, added manufacturing techniques, added trade routes, more labor) (Bunker, 2007; Gutowski et al., 2009; Strumsky et al., 2010). The major point here, however, is that the recognition of this political-ecological condition of solar PV technology depends upon how the boundary is drawn.

The vision and the reality of the solar future

The above calculations indicate that each country's solar PV aspiration makes demands on land that far exceeds the extent of the solar modules themselves. In each case, this demand is so large that the geographical territory within each country's borders is insufficient to accommodate the envisioned solar PV development. From this, we could conclude that a net import of resources from elsewhere is a necessary condition for successfully realizing large-scale solar PV visions. Such environmental load displacement, as we saw in the previous chapter, occurs through an ecologically unequal exchange that follows from relative price differences in the world economy. Such an exchange is ultimately upheld through disproportionate distribution of dismal working conditions, negligent environmental regulations, and a copious burning of fossil energy (notably coal) in the world economy. Under these conditions and assumptions, the large-scale solar PV projects of China, the US, Germany, India, and Italy all necessitate a highly politicized world division of labor as much as they necessitate polysilicon, engineers, electrical components, or direct sunshine. In this view, a world division of labor is an integral part of the new metabolism as far as it is based on large-scale construction of solar PV parks.

But do the solar PV projects analyzed above really necessitate such a world division of labor, as opposed to strongly encouraging it? What I have demonstrated is that large-scale solar PV projects necessitate a biophysical subsidy, as was concluded by Georgescu-Roegen already 40 years ago:

> The truth is that any present recipe for the direct use of solar energy is a 'parasite,' as it were, of the current technology, based mainly on fossil fuels. All the necessary equipment (including the collectors) are produced by recipes based on sources of energy other than the sun's. And it goes without saying that, like all parasites, any solar technology based on the present feasible recipes would subsist only as long as its "host" survives (Georgescu-Roegen, 1978: 19).

Several points in this quote are worth considering. First, as we saw in Chapter 3, the Industrial Revolution was successful precisely because feeding coal to machines provided access to copious amounts of "buried sunshine," ultimately representing large amounts of buried land. The fact that solar PV technology requires large surface areas could therefore theoretically be "solved" with a fossil energy subsidy. A country could theoretically develop a significant solar PV infrastructure if it was constructed with the help of fossil energy. Indeed, this seems to be the primary form of subsidy in modern Chinese manufacturing today. Also, in both China's and India's PV pathways, fossil energy consumption is projected to increase. Considering this, the large-scale development of solar PV technology does not strictly necessitate a world division of labor, but encourages it in cases where fossil energy is scarce or otherwise considered undesirable, i.e., it embodies a "weak inherent politics."

This conclusion, however, overlooks the fundamental political ecology of fossil energy in the world economy. To think that large-scale solar PV projects merely *encourage* a world division of labor by necessitating fossil energy is to forget that the access, use, and pollution of fossil fuels are politically charged processes intertwined with a world division of labor (Mitchell, 2011; Schaffartzik et al., 2014). Since solar PV modules show the greatest promise among the low-carbon alternatives when it comes to power density$_{st}$, it is unlikely that the biophysical subsidy could come from other renewable energy technologies. Seen from this perspective, large-scale solar PV projects embody a "strong inherent politics" because they necessitate a biophysical subsidy that could be granted through ecologically unequal exchange or through continued extraction and burning of fossil fuels. This means that large-scale solar PV development necessitates a world relation of power through which resources are unevenly distributed.

This leaves us with two technological boundaries with two distinct implications for interpreting the ontology of solar PV technology and the emerging metabolism. Considering this book's research question, which boundary is more realistic for assessing solar PV technology as an option for a socially just and ecologically sustainable world?

In the introductory paragraph to this chapter, both the Enclosure acts in Britain and the Valladolid debate in Spain justified social relations of power (capitalism and colonialism) by enforcing what cognitive scientists call "artificial" categories, i.e., categories that do not correspond to entities in the world. Out of the two technological boundaries, or categorizations, the narrower boundary of technology may seem to correspond more to the natural world than the broader boundary of technology as a system. After all, you might say, "here is the computer," and you might think of the computer as a technology. However, I would like to argue that the narrow definition is in fact more "artificial," because it excludes processes that are vital for the technology to exist in the world. There is after all no such thing as a functioning computer without a continuous electricity input, or a solar PV module without raw material extraction. The widely drawn boundary actually includes more of the biophysical world and should therefore be understood as a more realistic category.

It remains, however, that the broader category is not widely recognized in most people's everyday interaction with technological artifacts. The narrow definition of technology as an artifact remains more intuitive. This is a problem because it is precisely in this more intuitive and narrow understanding of technology that the feeling of a mystical technological power can take root and bloom (see 'fetishism' in the Glossary). Anthropologists have noted how modern technological faith is a type of magical thinking, where magic enters in the space between a collective wish and its fulfillment (Stivers, 1999). In the case of solar PV technology, magical thinking enters in instances where engineers and policymakers justify technological inefficiencies or environmental injustices with reference to a future progress in design. Recurring sentiments such as "if PV cells are redesigned, they can become more land-efficient in the future" or "if the commodity chain is altered, the technology can become more just" may in fact be expressions of such magical thinking. Such views are ultimately contingent on understanding solar PV modules as artifacts separate from the wider processes that generate them, rather than seeing these processes as necessary for their existence. Such views overlook the fundamentally material – and so social – ontology of technology (see Chapter 2). In effect, technological faith may implicitly and unintentionally justify social relations of power that are as much part of the technology as any material component that cannot be engineered away.

Similar to the Enclosure Acts and the Valladolid debate, these relations are encouraged and upheld through a cultural category that may seem correct, but which hides relations of power. With this in mind, it seems that the broader categorization of solar PV technology as intertwined with a world economy is more realistic if the aim is to establish a socially just and ecologically sustainable world. Given that the word "technology" is a fetishization of the complex sociotechnical systems emerging at the time of the Industrial Revolution, it is only fitting that the word is defetishized in the transformation away from fossil fuels. That said, it is crucial to remember that the conventional notion of power density$_{st}$ is the dominant boundary today, regardless of whether it is in some way irrational or problematic.

What does it mean to say that large-scale solar PV projects embody a strong inherent politics? Why does it matter and to whom? To understand the consequences of this, we can begin by recognizing that the strong politics informed by the extended boundary leads to a recognition of the incongruence between the proposed visions and the implications of their actual fulfillment. As we have seen, the contrast between the conception of technology and what technology is can unwittingly legitimize social relations of power and further exploitation of the natural world. In the case of China, the vision is to inaugurate an ecological civilization in which the world can flourish. Simultaneously, China's massive Belt and Road initiative is likely to draw on resources from throughout the world in a manner analogous to the imperial colonial era of the 1800s. For the Biden administration in the US, the vision is to maintain its independence from the growing economic might of China while catching up to, and surpassing the rest of the world, in climate policy. Simultaneously, the US energy independence is – and likely will remain – highly dependent on low-cost solar imports from Chinese corporations manufacturing at very low prices. India, as we have seen, is seeking to become energy independent by turning to large-scale solar PV projects. Simultaneously, India's vision can only be fulfilled via international collaborations that facilitate an import of vital raw materials, embodied labor, and embodied land from countries such as China, Thailand, and Indonesia. The German government is planning to phase out its domestic coal production by 2038 and plans to become more energy independent through the installation of renewable energy technologies. Simultaneously, this will necessitate a net import of resources from countries in the world economy that are currently expanding domestic coal production. Italy, meanwhile, is seeking higher energy independence and decarbonization. Simultaneously, it is seeking to achieve this through participation in a historically unprecedented infrastructure initiative aimed at expanding international exchange relations. In each case, admirable visions of an environmentally benign post-carbon transition are counteracted in their practical application.

The important consequence of this inconsistency between vision ("the map") and practical application ("the territory") is that the unfolding of the new metabolism is allowed to add to the social-ecological problems of the old, purely fossil-based, regime. The twist, however, is that the dismal working conditions, negligent environmental regulations, and copious burning of fossil energy can now be glossed over with reference to a progress in the development of renewable energy technologies. Ultimately, such progress may be an illusion maintained under narrowly drawn technological boundaries that fail to acknowledge the global political implications of the physical inefficiencies. Meanwhile, as the world gets warmer and the EROI of fossil energy carriers start slipping down the net energy cliff, the real risk is that different social groups start blaming each other for the atrocities and catastrophes of the world based on perceived differences. The resulting conflicts may in turn lead to justifications for further resource appropriation. To avoid this, it is highly relevant to align modern solar PV visions with the ecological reality of the world by recognizing what technology actually is.

Notes

1 In a substantial review (N = 11,500 screened, n = 835 analyzed) of "decoupling," Wiedenhofer et al. (2020) concluded that as few as 8% of all studies included consumption-based environmental loads occurring outside the national boundary.
2 This telos is generated by the historically developed relations of production and upheld by the social aspirations of powerful actors. Today, it pertains to the drive for infinite capital accumulation (see Chapter 3).
3 ANT scholars, such as Latour (1988), carried this position to an extreme, suggesting that technological artifacts themselves have agency by virtue of their consequential interaction with humans.
4 Winner defines the political as "arrangements of power and authority in human associations as well as the activities that take place within those arrangements" (1980).
5 A relation of power is broadly understood as a situation in which one group or individual has a disproportionate capacity to form decisions or carry out actions affecting another group or individual. Power, as the capacity to form decisions and carry out actions, is both social and physical (Russell et al., 2011).
6 Given the 20–30-year life span of solar PV modules, this situation is likely a continuing problem if the energy system reliant upon solar PV technology should become mature under capitalism.
7 Understood as global differences in wages, rent, and pricing of natural resources.
8 Further considerations on how to calculate power density extended is available in appendix B.
9 This is the percentage represented by the *added* solar PV capacity.
10 China's total surface area is 9,597,000 km^2.
11 The total surface area of the US is 9,826,675 km^2.
12 This is the sum of the indirect land requirements of labor and capital divided by the average life span of the solar PV modules (25 years) (see Table 5.9).
13 India's total surface area is 3,287,000 km^2.
14 Italy's total land surface area is 391,338 km^2.

References

AER. 2018. Stromsektor 2030 – Energiewirtschaftliche Auswirkungen von 65%Erneuerbare Energien und einer Reduktion der Kohleverstromung imEinklang mit den Sektorzielen des Klimaschutzplans. Kurzstudie.

AGEB. 2020. *Arbeitsgemeinschaft Energiebilanzenlegt Bericht zum Energieverbrauch 2019 vor.* Nr. 02.

Andersen, Otto. 2013. *Unintended consequences of renewable energy: Problems to be solved.* London, England: Springer.

Bang, Guri. 2021. The United States: Conditions for accelerating decarbonisation in a politically divided country. *International Environmental Agreements: Politics, Law and Economics* 21: 43–58. https://doi.org/10.1007/s10784-021-09530-x.

Bellini, Emiliano. 2020. China entering post-FIT era with solid prospects. *PV Magazine.* Published 17-06-2020. https://www.pv-magazine.com/2020/06/17/china-entering-post-fit-era-with-solid-prospects/. Accessed 23-09-2020.

Bhandari, Khagendra P., Jennifer M. Collier, Randy J. Ellingson, and Dfne S. Apul. 2015. Energy payback time (EPBT) and energy return on energy invested (EROI) of solar photovoltaic systems: A systematic review and meta-analysis. *Renewable and Sustainable Energy Reviews* 47: 133–141. https://doi.org/10.1016/j.rser.2015.02.057.

Bijker, Wiebe, Thomas Hughes, and Trevor Pinch, eds. 1989. *The social construction of technological systems: New directions in the sociology and history of technology.* Cambridge, MA and London, England: The MIT Press.

BMUB. 2016. *Climate action plan 2050: Principles and Goals of the German government's climate policy.* Berlin, Germany: Druck- und Verlagshaus Zarbock GmbH and Co. KG.

BMWI. 2020. Oil imports and crude oil production in Germany. *Federal Ministry of Economics Affairs and Energy.* https://www.bmwi.de/Redaktion/EN/Artikel/Energy/petroleum-oil-imports-and-crude-oil-productions-in-germany.html. Accessed 24-09-2020.

Bonds, Eric, and Liam Downey. 2012. "Green" technology and ecologically unequal exchange: The environmental and social consequences of ecological modernization in the world-system. *American Sociological Association* 18(2): 167–186. https://doi.org/10.5195/jwsr.2012.482.

BP. 2019a. *BP statistical review – 2019: China's energy market in 2018.* London: BP energy economics.

BP. 2019b. *BP statistical review – 2019: India's energy market in 2018.* London: BP energy economics.

Bruninx, Kenneth, Darin Madzharov, Erik Delarue, and William D'haeseleer. 2013. Impact of the German nuclear phase-out on Europe's electricity generation—A comprehensive study. *Energy Policy* 60: 251–261. https://doi.org/10.1016/j.enpol.2013.05.026.

Bunker, Stephen G. 2007. Natural values and the physical inevitability of uneven development under capitalism. In *Rethinking environmental history: World-system history and global environmental change*, edited by Alf Hornborg, J. R. McNeill, and Joan Martinez-Alier, 239–358. Lanham, MD and New York, NY: AltaMira.

Capellán-Pérez, Iñigo, Carlos de Castro, and Iñaki Arto. 2017. Assessing vulnerabilities and limits in the transition to renewable energies: Land requirements under 100% solar energy scenarios. *Renewable and Sustainable Energy Reviews* 77: 760–782. https://doi.org/10.1016/j.rser.2017.03.137.

Capellán-Pérez, Iñigo, Carlos de Castro, and Luis Javier Miguel González. 2019. Dynamic energy return on energy investment (EROI) and material requirements in scenarios of global transition to renewable energies. *Energy Strategy Reviews* 29: 100399. https://doi.org/10.1016/j.esr.2019.100399.

Carbajales-Dale, Michael, Marco Raugei, Vasilis Fthenakis, and Charles Barnhart. 2015. Energy return on investment (EROI) of solar PV: An attempt at a reconciliation. *Proceedings of the IEEE* 103(7): 995–999.

CAT. 2020a. China. *Climate action tracker.* https://climateactiontracker.org/countries/china/. Accessed 24-09-2020.

CAT. 2020b. USA. *Climate action tracker.* https://climateactiontracker.org/countries/usa/. Accessed 09-07-2022.

CAT. 2020c. Germany. *Climate action tracker.* https://climateactiontracker.org/countries/germany/. Accessed 24-09-2020.

CAT. 2020d. India. *Climate action tracker.* https://climateactiontracker.org/countries/india/. Accessed 27-09-2020.

CAT. 2020e. EU. *Climate action tracker.* https://climateactiontracker.org/countries/eu/. Accessed 27-09-2020.

Chacko, Priya. 2015. The new geo-economics of a "rising" India: State transformation and the recasting of foreign policy. *Journal of Contemporary Asia* 45(2): 236–344. https://doi.org/10.1080/00472336.2014.948902.

Chikkatur, Ananth P., and Ambuj D. Sagar. 2009. Rethinking India's coal-power technology trajectory. *Economics and Political Weekly* 44(46): 53–58.

Chiriboga, Gonzalo, Andrés De La Rosa, Camila Molina, Stefany Velarde, and Ghem Carvajal C. 2020. Energy return on investment (EROI) and life cycle analysis (LCA) of biofuels in Ecuador. *Heliyon* 6: e04213. https://doi.org/10.1016/j.heliyon.2020.e04213.

Cleveland, Cutler J., Charles A. S. Hall, Robert A. Herendeen, Nathan Hagens, Robert Costanza, Kenneth Mulder, Lee Lynd, et al. 2006. Energy returns on ethanol production. *Science* 312(5781): 1746–1748.

Coratella, Teresa. 2020. The European green deal: A political opportunity for Italy. *European Council on Foreign Relations.* Published 17-02-2020. https://www.ecfr.eu/article/commentary_the_european_green_deal_a_political_opportunity_for_italy. Accessed 24-09-2020.

Dauvergne, Peter, and Kate J. Neville. 2010. Forests, food, and fuels in the tropics: The uneven social ecological consequences of the emerging political economy of biofuels. *Journal of Peasant Studies* 37(4): 631–660. https://doi.org/10.1080/03066150.2010.512451.

de Castro, Carlos, and Iñigo Capellán-Pérez. 2020. Standard, point of use, and extended energy return on energy invested (EROI) from comprehensive material requirements of present global wind, solar, and hydro power technologies. *Energies* 13: 3036. https://doi.org/10.3390/en13123036.

de Castro, Carlos, Margarita Mediavilla, Luis Javier Miguel, and Fernando Frechoso. 2013. Global solar electric potential: A review of their technical and sustainable limits. *Renewable and Sustainable Energy Reviews* 28: 824–835. https://doi.org/10.1016/j.rser.2013.08.040.

DOE. 2021. Solar futures study. Published 09-2021. https://www.energy.gov/sites/default/files/2021-09/Solar%20Futures%20Study.pdf.

DOE. 2022. Solar manufacturing. *Office of Energy Efficiency and Renewable Energy.* https://www.energy.gov/eere/solar/solar-manufacturing. Accessed 09-07-2022.

Dusek, Val. 2006. *Philosophy of technology: An introduction.* Oxford, England: Blackwell.

Enkhardt, Sandra. 2019a. Germany lifts cap on solar FIT program with new Climate Change Act. *PV Magazine.* Published 23-09-2019. https://www.pv-magazine.com/2019/09/23/germany-lifts-cap-on-solar-fit-program-in-climate-change-act/. Accessed 24-09-2020.

Enkhardt, Sandra. 2019b. German government wants 98 GW of solar by 2030. *PV Magazine.* Published 9-10-2019. https://www.pv-magazine.com/2019/10/09/german-government-wants-98-gw-of-solar-by-2030/. Accessed 13-04-2020.

ERI. 2015. *China 2050 high renewable energy penetration scenario and roadmap study.* Beijing: Energy Research Institute National Development and Reform Commission and Energy Foundation.

Farrell, Alexander E., Richard J. Plevin, Brian T. Turner, Andrew D. Jones, Michael O'Hare, and Daniel M. Kammen. 2006. Ethanol can contribute to energy and environmental goals. *Science* 311(5760): 506–508. https://doi.org/10.1126/science.1121416.

Feenberg, Andrew. 1991. *Critical theory of technology.* New York, NY: Oxford University Press.

Ferroni, Ferruccio, Alexander Guekos, and Robert J. Hopkirk. 2017. Further considerations to: Energy return on energy invested (ERoEI) for photovoltaic solar systems in regions of moderate insolation. *Energy Policy* 107: 498–505. https://doi.org/10.1016/j.enpol.2017.05.007.

Ferroni, Ferruccio, and Robert J. Hopkirk. 2016. Energy return on energy invested (ERoEI) for photovoltaic solar systems in regions of moderate insolation. *Energy Policy* 94: 336–344. https://doi.org/10.1016/j.enpol.2016.03.034.

Georgescu-Roegen, Nicholas. 1978. Technology assessment: The case of the direct use of solar energy. *Atlantic Economic Journal* 6: 15–21. https://doi.org/10.1007/BF02300267.

Giampietro, Mario, Sergio Ulgiati, and David Pimentel. 1997. Feasibility of large-scale biofuel production. *Bioscience* 48(9): 587–600. https://doi.org/10.2307/1313165.

Griffiths, Ieuan. 1986. The scramble for Africa: Inherited political boundaries. *The Geographical Journal* 152(2): 204–2016. https://doi.org/10.2307/634762.

Gutowski, Timothy G., Matthew S. Branham, Jeffrey B. Dahmus, Alissa J. Jones, Alexandre Thiriez, and Dusan P. Skulic. 2009. Thermodynamic analysis of resources used in manufacturing processes. *Environmental Science and Technology* 43: 1584–1590. https://doi.org/10.1021/es8016655.

Hall, Charles A. S., Jessica G. Lambert, and Stephen B. Balogh. 2014. EROI of different fuels and the implications for society. *Energy Policy* 64: 141–152. https://doi.org/10.1016/j.enpol.2013.05.049.

Hall, Charles A. S., Stephen Balogh, and David J. R. Murphy. 2009. What is the minimum EROI that a sustainable society must have? *Energies* 2: 25–47. https://doi.org/10.3390/en20100025.

Hall, Max. 2019. Belt and Road to bring 1 GW of solar to Italy. *PV magazine.* Published 01-10-2019. https://www.pv-magazine.com/2019/10/01/belt-and-road-to-bring-1-gw-of-solar-to-italy/. Accessed 02-10-2020.

Hamouchene, Hamza. 2020. Desertec: What went wrong. *EcoMENA.* Published 23-07-2020. https://www.ecomena.org/desertec/. Accessed 26-09-2020.

Heikkurinen, Pasi. 2019. Degrowth: A metamorphosis in being. *Environment and Planning E: Nature and Space* 2(3): 528–547. https://doi.org/10.1177/2514848618822511.

Hermele, Kenneth. 2014. *The appropriation of ecological space: Agrofuels, unequal exchange and environmental load displacement.* Abingdon, Oxfordshire and New York, NY: Routledge.

Hornborg, Alf, Gustav Cederlöf, and Andreas Roos. 2019. Has Cuba exposed the myth of "free" solar energy? Energy, space, and justice. *Environment and Planning E: Nature and Space* 2(4): 989–1008. https://doi.org/10.1177/2514848619863607.

Huber, Matthew. 2015. Theorizing energy geographies. *Geography Compass* 9(6): 1–12. https://doi.org/10.1111/gec3.12214.

Huber, Matthew T., and James McCarthy. 2017. Beyond the subterranean energy regime? Fuel, land use and the production of space. *Transactions of the Institute of British Geographers* 42(2): 655–668. https://doi.org/10.1111/tran.12182.

Hughes, Thomas. 1989. The evolution of large technological systems. In *The social construction of technological systems: New directions in the sociology and history of technology,* edited by Wiebe Bijker, Thomas Hughes, and Trevor Pinch, 51–82. Cambridge, MA and London, England: The MIT Press.

Hutchins, Mark. 2019. Italy's FuturaSun opens 500 MW module fab in China. *PV Magazine.* Published 01-07-2020. https://www.pv-magazine.com/2019/07/01/italys-futurasun-opens-500-mw-module-fab-in-china/. Accessed 24-09-2020.

IEA. 2020a. *Snapshot of global photovoltaic markets.* Report IEA-PVPS T1–37:2017. Paris, France: Photovoltaic Power Systems Programme of the International Energy Agency.

IEA. 2020b. Total energy supply (TES) by source, Italy 1990–2019. Published 08-11-2020. https://www.iea.org/data-and-statistics?country=ITALYandfuel=Energy%20supply-andindicator=TPESbySource. Accessed 09-12-2020.

Isenhour, Cindy. 2016. Unearthing human progress? Ecomodernism and contrasting definitions of technological progress in the Anthropocene. *Economic Anthropology* 3: 315–328. https://doi.org/10.1002/sea2.12063.

Isenhour, Cindy, and Kuishuang Feng. 2014. Decoupling and displaced emissions: On Swedish consumers, Chinese producers and policy to address the climate impact of consumption. *Journal of Cleaner Production* 134: 320–329. https://doi.org/10.1016/j.jclepro.2014.12.037.

Jiborn, Magnus, Astrid Kander, Viktoras Kulinois, Hana Nielsen, and Daniel D. Moran. 2018. Decoupling or delusion? Measuring emissions displacement in foreign trade. *Global Environmental Change* 49: 27–34. https://doi.org/10.1016/j.gloenvcha.2017.12.006.

Jorgenson, Andrew K., and Brett Clark. 2012. Are the economy and the environment decoupling? A comparative international study, 1960–2005. *American Journal of Sociology* 118(1): 1–44.

Kan, Siyi, Bin Chen, and Guiqian Chen. 2019. Worldwide energy use across global supply chains: decoupled from economic growth? *Applied Energy* 250: 1235–1245. https://doi.org/10.1016/j.apenergy.2019.05.104.

Korzybski, Alfred. 2000[1933]. *Science and sanity: An introduction to non-Aristotelian systems and general semantics.* New York, NY: Institute of General Semantics.

Kunze, Conrad, and Paul Lehmann. 2019. The myth of the dark side of the Energiewende. In *The European dimension of Germany's energy transition. Opportunities and conflicts,* edited by Erik Gawel, Sebastian Strunz, Paul Lehmann, and Alexandra Purkus, 255–263. New York, NY: Springer.

Lambert, Jessica G., Charles A. S. Hall, Stephen Balogh, Ajay Gupta, and Michelle Arnold. 2013. Energy, EROI and quality of life. *Energy Policy* 64: 153–167. https://doi.org/10.1016/j.enpol.2013.07.001.

Latour, Bruno. 1988. Mixing humans and nonhumans together: The sociology of a door-closer. *Social Problems* 35(3): 298–310.

MacKay, David J. C. 2009. *Sustainable energy: Without the hot air.* Cambridge, England: UIT.

May, Nadine. 2005. *Eco-balance of a solar electricity transmission from North Africa to Europe.* Braunschweig: Technical University of Braunschweig.

McCarthy, James. 2015. A socioecological fix to capitalist crisis and climate change? The possibilities and limits of renewable energy. *Environmental and Planning A* 47: 2485–2502. https://doi.org/10.1177/0308518X15602491.

Medin, Douglas L., and Evan Heit. 1999. Chapter 3–Categorization. In *Cognitive science,* edited by Benjamin Martin Bly, and David E. Rumelhart, 99–143. San Diego, CA: Academic Press.

Meza, Edgar. 2020. Controversial new coal plant in Datteln 4 test runs disrupt German power prices. *Clean Energy Wire.* Published 19-05-2020. https://www.cleanenergywire.org/news/controversial-new-coal-plant-datteln-4-disrupts-germanys-green-power-sector. Accessed 20-09-2020.

Michalopolous, Stelios, and Elias Papaioannou. 2016. The long-run effects of the scramble for Africa. *American Economic Review* 106(7): 1802–1848. https://doi.org/10.1257/aer.20131311.

Mitchell, Timothy. 2011. *Carbon democracy: Political power in the age of oil.* London, England and New York, NY: Verso.

Murphy, Caitlin, Trieu Mai, Yinong Sun, Paige Jadun, Matteo Muratori, Brent Nelson, and Ryan Jones. 2021. *Electrification futures study: Scenarios of power system evolution and infrastructure development for the United Stated.* Golved, CO: National Renewable Energy Laboratory. NREL/TP-6A20-72330. https://www.nrel.gov/docs/fy21osti/72330.pdf.

Murphy, David J., and Charles A. S. Hall. 2010. Year in review—EROI or energy return on (energy) invested. *Annals of the New York Academy of Sciences* 1185(1): 102–118. https://doi.org/10.1111/j.1749-6632.2009.05282.x.

Okumu, Wafula. 2010. Resources and border disputes in Eastern Africa. *Journal of Eastern African Studies* 4(2): 279–297. https://doi.org/10.1080/17531055.2010.487338.

Palmer, Graham, and Joshua Floyd. 2017. An exploration of divergence in EPBT and EROI for solar photovoltaics. *Biophysical Economics and Resource Quality* 2: 15. https://doi.org/10.1007/s41247-017-0033-0.

Pickard, William F. 2014. Energy return on energy invested (EROI): A quintessential but possibly inadequate metric for sustainability in a solar-powered world? *Proceedings of the IEEE* 102(2): 1118–1122. https://doi.org/10.1109/JPROC.2014.2332092.

Pimentel, David. 2003. Ethanol fuels: Energy balance, economics, and environmental impacts are negative. *Natural Resources Research* 12(2): 127–134. https://doi.org/10.1023/A:1024214812527.

Pimentel, David, and Tad W. Patzek. 2005. Ethanol production using corn, switchgrass, and wood: Biodiesel production using soybean and sunflower. *Natural Resources Research* 14(1): 65–76. https://doi.org/10.1007/s11053-005-4679-8.

Prieto, Pedro A., and Charles A. S. Hall. 2013. *Spain's photovoltaic revolution: The energy return on investment.* New York, NY: Springer.

Raugei, Marco. 2019. Net energy analysis must not compare apples and oranges. *Nature Energy* 4(2): 86–88. https://doi.org/10.1038/s41560-019-0327-0.

Raugei, Marco, Pere Fullana-i-Palmer, and Vasilis Fthenakis. 2012. The energy return on energy investment (EROI) of photovoltaics: Methodology and comparisons with fossil fuel life cycles. *Energy Policy* 45: 576–582. https://doi.org/10.1016/j.enpol.2012.03.008.

Raugei, Marco, Sgouris Sgouridis, David Murphy, Vasilis Fthenakis, Rolf Frischknecht, Christian Breyer, Ugo Bardi, et al. 2017. Energy return on energy invested (ERoEI) for photovoltaic solar systems in regions of moderate insolation: A comprehensive response. *Energy Policy* 102: 377–384. https://doi.org/10.1016/j.enpol.2016.12.042.

Ritchie, Hannah, and Max Roser. 2022. United States: Energy country profile. *Our World in Data.* https://ourworldindata.org/energy/country/united-states. Accessed 07-07-2022.

Ritchie, Hannah, Pablo Rosado, and Max Roser. 2017a. CO_2 and greenhouse gas emissions. *Our World in Data.* Published 05-2017. https://ourworldindata.org/co2-and-other-greenhouse-gas-emissions. Accessed 24-09-2020.

Ritchie, Hannah, Pablo Rosado, and Max Roser. 2017b. Fossil fuels. *Our World in Data.* https://ourworldindata.org/fossil-fuels. Accessed 27-09-2020.

Russell, Edmund, James Allison, Thomas Finger, John K. Brown, Brian Balogh, and W. Bernard Carlsson. 2011. The nature of power: Synthesizing the history of technology. *Technology and Culture* 52(2): 246–259.

Schaffartzik, Anke, Andreas Mayer, Simone Gingrich, Nina Eisenmenger, Christian Loy, and Fridolin Krausmann. 2014. The global metabolic transitions: Regional patterns and trends in global material flows, 1950–2010. *Global Environmental Change* 26: 87–97. https://doi.org/10.1016/j.gloenvcha.2014.03.013.

Sengupta, Somini. 2020. China, in pointed message to US, tightens its climate targets. *The New York Times*. Published 22-09-2020. https://www.nytimes.com/2020/09/22/climate/china-emissions.html. Accessed 25-09-2020.

Sers, Martin R., and Peter A. Victor. 2015. The energy-emissions trap. *Ecological Economics* 151: 10–21. https://doi.org/10.1016/j.ecolecon.2018.04.004.

Shaw, Vincent. 2020. China added almost 4 GW of solar in first quarter despite Covid-19. *PV Magazine*. Published 20-05-2020. https://www.pv-magazine.com/2020/05/20/china-added-almost-4-gw-of-solar-in-first-quarter-despite-covid-19/. Accessed 23-09-2020.

Shearer, Christine, Aiqun Yu, and Ted Nace. 2019. Out of step: China is driving the continued growth of the global coal fleet. Global Energy Monitor. Published 11-2019. https://globalenergymonitor.org/report/out-of-step-china-is-driving-the-continued-growth-of-the-global-coal-fleet/

Shearer, Christine, Robert Fofrich, and Steven J. Davis. 2017. Future CO_2 emissions and electricity generation from proposed coal-fired power plants in India. *Earth's Future* 5: 408–416.

Shiva, Vandana. 2008. *Soil not oil: Environmental justice in an age of climate crisis*. London, England: Zed Books.

Sinha, Dhruv. 2020. Why India is the new hotspot for renewable energy investors. *Indiaincgroup*. Published 18-05-2020. https://indiaincgroup.com/why-india-is-the-new-hotspot-for-renewable-energy-investors/. Accessed 23-09-2020.

Smil, Vaclav. 2006. 21st century energy: Some sobering thoughts. *OECD Observer* 258/259: 22–23.

Smil, Vaclav. 2015. *Power density: A key to understanding energy sources and uses*. Cambridge, MA: MIT Press.

Spencer, Thomas, Neshwin Rodrigues, Raghav Pachouri, Shubham Thakre, and G. Renjith. 2020. *Renewable power pathways: Modelling the integration of wind and solar in India by 2030*. New Delhi, India: The Energy and Resources Institute.

Stivers, Richard. 1999. *Technology as magic: The triumph of the irrational*. New York, NY: Continuum.

Stock, Ryan, and Trevor Birkenholtz. 2019a. The sun and the scythe: Energy dispossessions and the agrarian question of labor in solar parks. *The Journal of Peasant Studies* 48(5): 1–24. https://doi.org/10.1080/03066150.2019.1683002.

Stock, Ryan, and Trevor Birkenholtz. 2019b. Photons vs. firewood: Female (dis)empowerment by solar power in India. *Gender, Place and Culture* 27(1): 1628–1651. https://doi.org/10.1080/0966369X.2020.1811208.

Strumsky, Deborah, José Lobo, and Joseph Tainter. 2010. Complexity and the productivity of innovation. *Systems Research and Behavioral Science* 27(5): 496–509. https://doi.org/10.1002/sres.1057.

Susskind, Lawrence, Jungwoo Chun, Alexander Gant, Chelsea Hodgkins, Jessica Cohen, and Sarah Lohmar. 2022. Sources of opposition to renewable energy projects in the United States. *Energy Policy* 165: 112922. https://doi.org/10.1016/j.enpol.2022.112922.

Sylvia, Tim. 2022. Bidgen admin to pause new solar tariffs for two years. *PV Magazine*. Published 06-06-2022. https://pv-magazine-usa.com/2022/06/06/breaking-biden-admin-set-to-pause-new-solar-tariffs-for-two-years/. Accessed 07-07-2022.

The White House. 2022a. President Biden to highlight clean energy manufacturing and deployment investments that cut consumer costs, strengthen US energy sector, and create good-paying jobs. Published 28-02-2022. https://www.whitehouse.gov/briefing-room/statements-releases/2022/02/28/fact-sheet-president-biden-to-highlight-clean-energy-manufacturing-and-deployment-investments-that-cut-consumer-costs-strengthen-u-s-energy-sector-and-create-good-paying-jobs/. Accessed 09-07-2022.

The White House. 2022b. President Biden takes bold executive action to spur domestic clean energy manufacturing. Published 06-06-2022. https://www.whitehouse.gov/briefing-room/statements-releases/2022/06/06/fact-sheet-president-biden-takes-bold-executive-action-to-spur-domestic-clean-energy-manufacturing/. Accessed 09-07-2022.

US Department of State. 2021. *The long-term strategy of the United States: Pathways to net-zero greenhouse gas emissions by 2050*. Washington, DC: e United States Department of State and the United States Executive Office of the President.

Wagman, David. 2021. First reactions to Biden's US ban on solar imports from China's Xinjiang province. *PV Magazine*. Published 24-06-2021. https://www.pv-magazine.com/2021/06/24/first-reactions-to-bidens-us-ban-on-solar-imports-from-chinas-xinjiang-province/. Accessed 06-07-2022.

Weißbach, D., G. Ruprecht, A. Huke, K. Czerski, S. Gottlieb, and A. Hussein. 2013. Energy intensities, EROIs, and energy payback times of electricity generating power plants. *Energy* 52: 2010–2221. https://doi.org/10.1016/j.energy.2013.01.029.

Wettengel, Julian. 2020. Spelling out the coal exit—Germany's phase-out plan. *Clean Energy Wire*. Published 03-07-2020. https://www.cleanenergywire.org/factsheets/-spelling-out-coal-phase-out-germanys-exit-law-draft. Accessed 20-09-2020.

Wiedenhofer, Dominik, Doris Virág, Gerald Kalt, Barbara Plank, Jan Streeck, Melanie Pichler, Andreas Mayer, Fridolin Krausmann, Paul Brockway, and Anke Schaffartzik. 2020. A systematic review of the evidence on decoupling of FDP, resource use and GHG emissions, part I: Bibliometric and conceptual mapping. *Environmental Research Letters* 15(6): 063002. https://doi.org/10.1088/1748-9326/ab8429.

Winner, Langdon. 1980. Do artifacts have politics? *Daedalus* 109(1): 121–136. https://www.jstor.org/stable/20024652.

Yenneti, Komali, Rosie Day, and Oleg Golubchikov. 2016. Spatial justice and the land politics of renewables: Dispossessing vulnerable communities through solar energy mega-projects. *Geoforum* 76: 90–99. https://doi.org/10.1016/j.geoforum.2016.09.004.

6 The world and the solar module

This book set out with the hypothesis that solar photovoltaic (PV) technology is conventionally perceived in ways that are incongruent with the physical conditions of its existence. Through philosophical, historical, and empirical considerations of solar PV technology, I have now shown that the social and ecological conditions of its existence are indeed more complex than is generally assumed. In particular, I have shown how ecologically unequal exchange is a likely prerequisite for solar PV technology and how this condition is repeatedly overlooked by researchers, policymakers, corporations, and governments who are working toward a low-carbon transition. Thus, solar PV technology has been embraced without a full understanding of its global prerequisites. Metaphorically speaking, we see the mushroom but not the mycelium. In this chapter, I provide a summary of the conclusions and revisit the larger philosophical question of what technology ultimately *is*.

The world in the solar module

This book has shown that contemporary solar aspirations tend to require an ecologically unequal exchange through which embodied labor and matter-energy flow from one social group to another. This condition may be inherent in large-scale solar PV technology projects, such as they are now being pursued by industrialized nations. The fact that Germany appropriated a notable amount of embodied resources from China in order to realize its solar PV aspirations during the period of 2002–2018, is testimony to the significance of this relation. So is the fact that the solar aspirations of leading solar nations, including China, the US, Germany, India and Italy, require an amount of land that would make a complete low-carbon energy transition unfeasible without substantial biophysical subsidies. Such biophysical subsidies could be obtained either by continuing to extract and burn fossil fuels or by exchanging commodities at favorable prices, as Germany did during the crucial period of 2002–2018.[1,2]

This finding shows how fulfilling large-scale development of solar PV technology necessitates globally uneven relations as much as they necessitate polysilicon, electrical components, engineers, or direct sunshine. While solar PV modules may be employed with or without specific social or ecological goals in mind,

DOI: 10.4324/9781003292319-6

I have highlighted how the very existence of the modules may require social asymmetries in resource distribution. This includes the uneven valuation of labor, resources, and pollution in the world peripheries relative to the world core. Solar PV technology can thus be categorized as "authoritarian" in Mumford's sense and as "inherently political" in Winner's sense (Mumford, 1964; Winner, 1980). This implies that solar power is intertwined with global political economy just as much as fossil fuels. The difference is that the politics of fossil fuels pertains primarily to these fuels as energy carriers (coal, oil, gas), whereas the politics surrounding solar PV modules primarily pertains to the massive amount of material infrastructure necessary for capturing the dispersed and intermittent flow of direct sunshine.[3] Since these requirements stem from the physical circumstances of capturing large amounts of direct sunshine through technically sophisticated and materially voluminous artifacts, it is not likely that this condition could be transformed even under new relations of production. In Marxist terminology, the relations of production and the means of production are inseparable. Globally uneven resource transfers appear to be an integral part of the envisioned socio-ecological regime based on capturing large amounts of energy with solar PV technology.

This conclusion has consequences for whether further pursuits of solar PV technology contribute to a continuation, transcending, or reversal of the industrial regime (see Chapter 3). Conventional interpretations of nature, technology, history, and development anticipate that progress in solar PV technologies will contribute to a world in which climate change and global injustices are being mitigated, while industrialism and affluent high-energy lifestyles *continue* as normal. Such perspectives tend to overlook the fact that technological artifacts are contingent on both the natural world and the world economy. This neglect is most apparent in the common assertion that solar power is "free," "abundant," and problem-free (e.g., Schwartzman, 1996; Bastani, 2019; McKibben, 2019). The fact that solar PV projects require large capital investments and large amounts of land, energy, and material per unit of energy harnessed suffices to invalidate this assertion. Solar power is not free and abundant. Solar PV infrastructure comes at the expense of the natural world and emerges within a context of world inequalities, as shown by the fact that the drop in prices in solar PV modules over the last 20 years occurred because manufacturers took the initiative to exploit international wage and price differences.

A point raised by those who believe that solar PV technology presents a way to *transcend* the industrial regime is that the energy harnessed from solar power cannot be used to power the global transport system currently propelled by easily transportable and highly energy-dense energy carriers, such as oil (see Chapter 3). This is so because solar power, given its landscape-dependent character, cannot be used to transport manufacturing facilities to regions in the world where wages are low without great economic costs. This, so the argument goes, will put an end to a 200-year-old arbitrage in which transnational corporations have had the power to shape the demands and regulations of global manufacturing by threatening to move production facilities to other regions of the world. As this book shows, however, ending this exploitation of the world peripheries might in fact increase prices on solar PV modules and halt the rate of installation. Fulfilling global economic aspirations for solar power would then undermine the process of capital accumulation and the societal prerequisites for solar PV technology.

Importantly, *price differences* are the key determinant for the availability of solar PV modules in the world economy. Ecologically unequal exchange, whereby matter-energy flows from one region to another, will occur only if wages and prices on raw materials and carbon sinks are higher in one region than in another. In effect, if global transportation can be maintained, albeit in a more costly form, perhaps utilizing battery power and wind power, then the conditions for producing solar PV modules might still be upheld even in the absence of fossil fuels. The higher costs in transportation would not render trade impossible, but it would make the commodities traded less available for the many. In such a context, the larger the price differences in the world economy, the more affordable is solar PV modules in wealthier regions of the world. Therefore, in the absence of fossil fuels, it is likely that intensified global asymmetries are necessary for societies (or specific social groups within them) aspiring to maintain high-energy lifestyles. Thus, the question is not if a new socio-ecological regime is possible or not, but rather for whom and how many it will be available. The following section discusses what this says about solar aspirations based on the premise that solar PV technology is a feasible option for a socially just and ecologically sustainable world.

The solar module in the world

I now wish to reflect on some of the issues I raised in the introduction. Let us recall that ETC's solar park in Katrineholm was a showcase for demonstrating how easy and affordable it is to "save the world" by purchasing solar PV cells and organizing in citizen energy cooperatives. This combination of solar power and local democracy, so the argument went, had already been demonstrated in Germany. We now know that a reason why a large share of Germany's electricity could by generated by democratically organized renewable energy cooperatives was because Germany benefited from uneven resource transfers in the world economy. It is peculiar how easy it was to spend a whole day in the solar park without encountering any information about this. As I walked around the park, my attention was directed to the various types of solar cells and to information about their technical properties. It occurred to me only much later that the solar modules were exhibited as being *in* the world but not *of* the world. The solar modules had undergone a process of fetishization through which knowledge about the early stages of their commodity chain had been erased and replaced by posters focusing on their features as technical artifacts. Following Marx's definition of fetishism, this was "nothing but the definite social relation between men themselves which [assumed] here, for them, the fantastic form of a relation between things" (1990[1867]: 165).

The closest to a mention of global asymmetries was a paragraph on one of the information signs by the park's row of thin film solar panels. With a headline in capital letters saying "thin film was the melody of the future," the sign reads:

A few years ago, thin film was thought to become the dominant solar technology. These 7 kW German thin film panels are typical: glass panels with a micro-thin film inside which generates electricity. Many people thought that the "old technology" - silicon solar cells - would be replaced by this type. But instead, *silicon solar cells dominate the world market. These cells have simply become cheaper.* Now they are even made with all-black backgrounds so that they look like thin film. Thin film needs about 25–30 percent more surface area than silicon solar cells to produce the same amount of electricity. Of course, many thin film producers say that it is a better product because it can handle clouds better than silicon. We have tried this in the park and reality is our judge. We have not been able to see any difference between thin film and silicon solar cells for the whole year. Both technologies are equally good (my translation and emphasis).

By acknowledging the existence of a world market, the author of the sign came close to recognizing the global asymmetries that I have examined. However, nothing on the sign suggested anything as politically charged as Chinese low-wage labor, fossil-propelled manufacturing, deregulated mineral extraction, or repression of local resistance. Except for David's meditation on Chinese wages, nothing in the park suggested that the words "cells have simply become cheaper" actually referred to a world division of labor and an uneven exchange of resources and environmental harms.

The relation between global asymmetries and PV prices is sometimes acknowledged. When it is acknowledged, however, it is also typically downplayed or ignored. In Chapter 4, we saw an unwillingness to consider the world division of labor as a necessary condition for the solar boom. In Chapter 5, we saw how the world division of labor was downplayed by drawing a narrow boundary around what constitutes solar PV technology. The unwillingness to integrate global asymmetries as a prerequisite for solar PV modules is obvious in the case of ETC's solar park, where the issue is acknowledged but not communicated. All this suggests that some people who are now pursuing solar PV technology are aware of the relation between the world division of labor and prices of solar modules, but knowingly disregard it. Why? I can see at least four interrelated explanations for why this fetishization of PV technology occurs.

a　*Alienation.* The fetishized perspective reflects an everyday experience of technologies wherein the biophysical past of the artifact is seldom seen as relevant. The conventional perspective on solar PV technology is encouraged by the fact that PV modules are commodities produced in a highly complex world economy. Thus, the world economic system itself encourages a fetishized perspective as an aspect of alienation. For this reason, people may be aware of the relation between global asymmetries and technology but struggle to integrate the connection in their everyday lives.

b　*Power.* Fetishization may be a strategy for legitimizing the pursuit of power. We should recall that ETC's park was founded on the aspiration to become a political force of consequence. Similarly, the leading solar nations' ambitions to install massive amounts of solar PV modules were regularly justified with reference to greater national energy autonomy (Chapter 5). In the case of China, solar PV development was also associated with the ambition to become an "ecological civilization" with great influence over the trajectory of the world economy. By omitting the negative effects of solar PV development, these pursuits for social-physical power appear as more legitimate.

c   *Ideology.* Recognizing the global social conditions of technology clearly challenges the conventional perception of what technology *is*. Conventional solar visions typically assume that solar PV technologies can be employed to solve problems. However, because technologies necessitate global asymmetries, they tend to solve problems for some at the expense of others and are better understood as entropy displacers (see Chapter 2). Ultimately, this understanding of technology does not fit into the worldview according to which solar PV technology is predestined to encourage socially just and environmentally friendly ways of living. As ideology, fetishism may derive from a deep-seated desire or socially encouraged conviction that "prevents, [or] renders even unnoticeable, contrary evidence or argument" (Pippin, 1994: 96). This is reminiscent of Heidegger's notion of "Enframing" as something which prevents a person from entering into "a more original revealing [i.e., how things come to be] and hence to experience the call of a more primal truth" (1977[1954]: 28).

d   *Denial.* The notion that global inequality is a prerequisite for solar PV modules may be threatening to people who perceive themselves as morally good people advocating PV technology for good reasons. In such a context, adopting the fetishized perspective may be a strategy to protect oneself from threatening information. Cohen's (2001: 51) definition of denial as "the maintenance of social worlds in which an undesirable situation … is unrecognized, ignored or made to seem normal" seems to apply to the case of solar PV fetishism. An "interpretative denial," in which threatening information is distorted, might be applicable to some of the explanations of the solar boom and the physical inefficiencies of solar technology (Wullenkord and Reese, 2021). In turn, an "implicatory denial" might apply to the case of ETC, in which a fact was recognized but "not integrated into everyday life or translated into action" (ibid. 6).

The wider effect of this solar technology fetishism is that it obscures the global asymmetries of the emerging energy regime.

A few years ago, one study concluded that ExxonMobil Corporation had purposefully misled the public regarding climate change and its existential implications (Supran and Oreskes, 2020). ExxonMobil, it turns out, had shared scientifically correct assessments of climate change internally but communicated skepticism and denial in public media in order to avoid stranded fossil assets. The question is whether the strategic omission of the global distributive dimension of solar PV technology may one day be compared to ExxonMobil's strategic denial of climate change. It is arguably too early to predict the wider biogeochemical implications of the new energy regime and its effects on ecosystems (cf. Lenton et al., 2016; Rehbein et al., 2020). Solar-generated electricity still comprises only 1% of global primary energy consumption (OWD, 2022). However, with ambitious efforts to massively increase the share of solar power in the world economy, we had better consider its prerequisites already today, so that we can envision and put into practice the new energy regime with full awareness of the conditions of pivotal technologies such as solar PVs.

A critical ecological ontology of solar technology

To discuss the ontology of technology, I have found it helpful to start with the analogy of a tree. In this analogy, the leaf is to the solar module what the root system is to the world economy. A leaf and a solar module can both harness the energy in direct sunshine but neither of them can exist without a root system or a world economy, respectively. To omit the biophysical past of solar PV modules as an essential aspect of what technology is (i.e., fetishizing it), is similar to omitting the root system from the conception of what constitutes a tree. The analogy also highlights how the root system and the world economy are both hidden from view, even if we know that they both exist. By demonstrating how solar PV modules necessitate ecologically unequal exchange, I have highlighted how a specific root system is necessary for the existence of the leaf. To be more concrete, I have demonstrated how solar modules and global asymmetries in the world economy are two aspects of an inseparable whole, i.e., solar PV technology. These are inseparable in a way that corresponds to the inseparability of the relation between the leaves and the roots of a tree. These respective parts can be analytically but not physically separated without compromising the survival of the tree.

Is it correct, then, to compare technology with an organism? I would argue that technology is better understood as an organ, rather than the organism itself (cf. Ellul, 1964). A technology made up of leaves and roots is not comparable to an organism, but rather a metabolic strategy for reproducing it. In a similar sense, we can think of technologies as the historically developed metabolic organs of society. This, I would argue, is what Marx was reaching for when he asserted that technology is the human equivalent to "the formation of the organs of plants and animals, which serve as the instruments of production for sustaining their life" (1990[1867]: 493, footnote 4). Marx had a social rather than individualist understanding of humans, which means that he understood technologies as the social organs by which human collectives interact with their environment. As demonstrated by Lewis Mumford (1954, 1964, 1967), such social organs – which he called "megamachines" – have existed for a long time in human history. However, as we have seen, it was only with the rise and acceleration of the industrial regime that the modern social organs were fetishized and labeled technology. Modern technologies – simultaneously cultural fetishes and material realities – are organs of the industrial metabolism, which implicate a world division of labor in the same sense that the leaf implicates a root system.

This means that the uneven distribution of resources associated with numerous environmental justice conflicts in the solar PV commodity chain cannot be dissolved, merely shifted around among social groups, biogeochemical cycles, and ecosystems. This represents the core of the issue concerning shifting away from fossil fuels by means of advanced technologies, since the underlying matter-energy throughput will merely be transformed in kind, not magnitude. As far as climate change is mitigated by way of solar PV technology without addressing this throughput, the risk is that climate change will be replaced by another eco-existential concern of equal magnitude (albeit of a different kind).

This forces us to seriously consider to what degree a long-term sustainable relation to the biosphere can be reached through an endless expansion of the technosphere, as implied in proposals for "green" growth or various "green" technologies, and to what degree it can only be reached by a progressive degrowth with attention to well-being, justice, and ecological limits. Rapid and extensive efforts to install solar PV technology and other renewables may encourage a world with less greenhouse gas emissions. In the process, however, these efforts generate new or intensify old problems in the Earth system that affected communities, activists, and researchers are only starting to understand (see Chapter 1). It is abundantly clear that fossil fuels are finite and non-sustainable energy carriers that the world's actors must immediately stop burning in order to halt the detrimental effects of climate change (IPCC, 2014, 2018). The pressing question is not whether the ecological effects of fossil fuels present a better or worse ecological situation than the effects of solar power. Rather, the question is whether a socially just and ecologically sustainable world is currently being pursued through solar power and other renewables. At the very least, I have shown that this is a more complex question than is generally assumed. Large-scale solar PV development, in particular, may generate problems that in the long term may be as detrimental to the biosphere as climate change.

Notes

1   Theoretically, such a subsidy could also be obtained through tribute payments or by appropriating resources through plunder (e.g., Wallerstein, 2004: 28). Since solar PV technology is the most land efficient renewable energy technology, it is unlikely that a biophysical subsidy could come from other renewable energy technologies such as wind, water, or geothermal.
2   Even prior to this period, a division of labor was evident in Germany's solar PV industry, but it was then confined to relative price differences within the nation, i.e., between the western and eastern parts of the country.
3   To be sure, installation and operation of fossil infrastructure is also a significant and highly politicized issue. In comparison to solar PV technology, however, the infrastructure necessary for fossil extraction is relatively small per unit of energy harnessed (Smil, 2015).

References

Bastani, Aaron. 2019. *Fully automated luxury communism*. London, England and New York, NY: Verso.

Cohen, Stanley. 2001. *States of denial: Knowing about atrocities and suffering*. Cambridge, England: Polity Press.

Ellul, Jacques. 1964. *The technological society*. New York, NY: Vintage.

Heidegger, Martin. 1977[1954]. *The question concerning technology and other essays*. New York, NY and London, England: Garland Publishing.

IPCC. 2014. *Climate Change 2014: Synthesis report*. Contribution of working groups I, II and III to the fifth assessment report of the intergovernmental panel on climate change. IPCC, Geneva, Switzerland.

IPCC. 2018. Summary for policymakers. In *Global warming of 1.5°C. An IPCC special report on the impacts of global warming of 1.5°C above pre-industrial levels and related global greenhouse gas emission pathways, in the context of strengthening the global response to the threat of climate change, sustainable development and efforts to eradicate poverty*, edited by Masson-Delmotte, V., P Zhai, H.O. Pörtner, D. Roberts, J. Skea, P.R. Shukla, A. Pirani, et al. 3–24. Cambridge and New York: Cambridge University Press. https://doi.org/10.1017/9781009157940.00.

Lenton, M. Timothy, Peter-Paul Pichler, and Helga Weisz. 2016. Revolutions in energy input and material cycling in earth history and human history. *Earth System Dynamics* 7: 353–370. https://doi.org/10.5194/esd-7-353-2016.

Marx, Karl. 1990[1867]. *Capital: A critique of political economy*. London, England: Penguin.

McKibben, Bill. 2019. If the world ran on sun, it wouldn't fight over oil. *The Guardian*. Published 18-09-2019. https://www.theguardian.com/commentisfree/2019/sep/18/climate-crisis-oil-war-iraq-saudi-attack-green-energy?fbclid=IwAR0prXxroXRD83FDQgbaNv8METigaalLFYvf89JcjZB0JLctWqr-bPSYgM4. Accessed 30-11-2020.

Mumford, Lewis. 1954. *Technics and civilization*. London, England: Routledge and Kegan Paul Ltd.

Mumford, Lewis. 1964. Authoritarian and democratic technics. *Technology and Culture* 5(1): 1–8. https://doi.org/10.2307/3101118.

Mumford, Lewis. 1967. *The myth of the machine: Technics and human development*. New York: Harcourt Brace Jovanovich.

OWD (Our World in Data). 2022. Global primary energy consumption by source. https://ourworldindata.org/grapher/global-energy-consumption-source?country=~OWID_WRL. Accessed 26-07-2022.

Pippin, Robert R. 1994. On the notion of technology as ideology: Prospects. In *Technology, pessimism, and postmodernism*, edited by Yaron Ezrahi, Everett Mendelsohn, and Howard Segal, 99–113. Dordrecht, the Netherlands: Springer.

Rehbein, Jose A., James E. M. Watson, Joe L. Lane, Laura J. Sonter, Oscar Venter, Scott C. Atkinson, and James R. Allan. 2020. Renewable energy development threatens many globally important biodiversity areas. *Global Change Biology* 26(5): 3040–3051. https://doi.org/10.1111/gcb.15067.

Schwartzman, David. 1996. Solar communism. *Science and Society* 60(3): 307–331.

Smil, Vaclav. 2015. *Power density: A key to understanding energy sources and uses*. Cambridge, MA: MIT Press.

Supran, Geoffrey, and Naomi Oreskes. 2020. Assessing ExxonMobil's climate change communications (1977–2014). *Environmental Research Letters* 12: 084019. https://doi.org/10.1088/1748-9326/ab89d5.

Wallerstein, Immanuel. 2004. *World-systems analysis: An introduction.* Durham, NC and London, England: Duke University Press.

Winner, Langdon. 1980. Do artifacts have politics? *Daedalus* 109(1): 121–136.

Wullenkord, Marlis, and Gerhard Reese. 2021. Avoidance, rationalization, and denial: Defensive self-protection in the face of climate change negatively predicts pre-environmental behavior. *Journal of Environmental Psychology* 77: 101683. https://doi.org/10.1016/j.jenvp.2021.101683.

7 Solutions beyond solar illusions

The message of this book is a message of disillusionment. The upside of disillusionment, which often carries a negative connotation, is a deeper appreciation and understanding of how tacit assumptions influence human behavior. In Chapter 1, I formulated the hypothesis that "the conception of solar PV technology contradicts the actual conditions of its existence." I have now unpacked the contrast between "the vision and reality" of solar photovoltaic technology for the purpose of understanding its true transformative potential and come to the conclusion that solar PV technology may owe its transformative potential to a globally uneven distribution of resources. This conclusion, I hope, is illuminating. It invites us to consider how the conception of solar PV technology – and the category of technology more generally – hides an important dimension of what solar PV technology *is*. The modern notion of technology is not only problematic from the standpoint of global environmental justice, but also instrumental for legitimizing a continued exploitation of nature. Ironically then, yet true to its purpose, the philosophical materialism at the foundation of this book leads to the conclusion that we ought to *rethink* the category of solar technology. By the end of the previous chapter (Chapter 6), I demonstrated how such a rethinking might look. Disillusionment and rethinking is arguably a necessary step for attempts at a social-metabolic transformation toward greater global environmental justice.

While disillusionment is a noble task, it is not a happy one. Many people today, especially young people, are anxious and feel hopeless about the future. In a recent study, Hickman et al. (2021) surveyed the thoughts and feelings about climate change in 10,000 children and young people from around the world. The results showed that a majority was very or extremely worried about climate change. This climate anxiety also correlated with a feeling of betrayal. The anxiety of these young people reflect the very real tragedy of the ecological crisis and the worldwide betrayal of governments and corporations not living up to their much-advertised visions. Widespread climate anxiety is therefore not an illness to cure, but an emotion to acknowledge, communicate, and consciously act upon. Environmentalists have long recognized this condition and experience. As once expressed by Aldo Leopold (1993[1953]: 147),

DOI: 10.4324/9781003292319-7

One of the penalties of an ecological education is that one lives alone in a world of wounds. Much of the damage inflicted on land is quite invisible to laymen. An ecologist must either harden his shell and make believe that the consequences of science are none of his business, or he must be the doctor who sees the marks of death in a community that believes itself well and does not want to be told otherwise.

Older generations in the Global North, much like Leopold's "laymen," seem less sensitive to the existential threat of the ecological crisis. Their childhood was very different from the childhood of today's young people who are growing up in a world of more frequent extreme weather events, rising relative living costs, increased violence and constant reminders of the biological state of the planet (to mention only a few). More importantly, not everyone can or want to adopt the role of "the doctor who sees the marks of death" in their community and in their environment. Such a role is demanding and often requires a high level of psychological need satisfaction (Wullenkord, 2020). Taking on this role may also be prevented by a form of action paralysis when facing the overwhelming magnitude of the problem. In such cases, the "ecological education" of "a world of wounds" becomes practically unhelpful. For these reasons, it is my conviction that the understanding of the state of the natural world and the improbability of future human development and justice based on current technological recipes must be countered with *realistic visions of desirable futures.*

What I have shown throughout this book is that solar visions are often desirable but not realistic. This does not mean that solar PV modules are impracticable and cannot be mass-produced and installed in increasing volumes (they currently are), merely that this will not achieve the purposes for which they were originally intended. For people who believe in the determinant force of technological progress and a linear course of history, this insight could be cognitively and ideologically threatening. The remedy to these obstacles, I assume, is to offer desirable solutions going beyond conventional solar visions. This involves offering something to live for beyond the conventional vision of a society in which an abundance of resources, supplied gratis through solar power and other renewables, will provide the individual with an increasing amount of commodities satisfying their every need. Much like this conventional vision of a materially affluent society, realistic visions also needs to be positive, fulfilling and attractive to the individual. The challenge is that they must be appealing enough that the individual is compelled to abandon the condition in which they define themselves as spectators of an already pre-destined history, in which the human enterprise unlocks the secret of limitless resources and leaves the planet.

In this chapter, I focus on the notion of "realistic envisioning" as a helpful tool for individuals, local communities, grassroots organizations, and perhaps even politicians to formulate realistic visions of desirable future societies. This is followed by an explanation and discussion on the ramifications of an "alternative solar technology" based on photosynthesizing organisms. Following the assertions of practical materialism (see Chapter 2), I thereafter discuss how realistic envisioning must be connected to contemporary social movements forming alliances for transformative change. In this sub-section, I introduce the notion of a "metabolic counter-regime," which provides the Polanyian "countermovement" with a proactive edge in relation to the social and environmental destruction of largely unregulated market forces exploiting global price differences.

Realistic envisioning

The realism that I here refer to pertains to physical reality. For a technology to be realistic, I therefore mean that it should be physically achievable and that such a technology can conceivably exist in the world despite nature's limits, boundaries, and laws. It does not, as is more common, refer to being politically viable or politically popular.[1] By realistic I also refer to a technology's capacity to live up to the socially articulated visions for which it was originally intended. All technologies are laden with the vision and the values motivating people to design it for certain purposes. In the case of solar PV technology, we have seen that such visions and values often refer to climate mitigation, energy independence, energy security, ecological sustainability, or social justice. While the first (physical) kind of realism takes ontological precedence over the second (teleological) definition, the two are interlinked. The vision and values given onto a technology cannot be achieved if that technology is not physically possible within a natural environment. Simultaneously, the technology loses its relevance to exist if it cannot live up to at least some desirable social visions or values. What I have shown in this book is that a particular social vision can be achieved even if the technology is physically near impossible within a given natural environment. The mechanism of ecologically unequal exchange allows for one people's vision to be achieved at the cost of increasing physical burdens among other environments and peoples. In the case of solar power, the injustices implicated in ecologically unequal exchange contradicts conventional visions of solar power, rendering it unrealistic as a global aspiration.

If by realism I refer to the physical, envisioning pertains to human culture and the capacity to imagine desirable futures. The increasingly popular concept of "socio-technical imaginaries" comes close to what I mean by envisioning. This concept is commonly attributed to Jasanoff and Kim's (2009) study on how the national political cultures of the US and South Korea informed divergent visions of the role of nuclear power. Defined as "collectively imagined forms of social life and social order reflected in the design and fulfillment of nation-specific scientific and/or technological projects" (ibid. 120), an imaginary is a type of social vision of a desirable future. As opposed to a socio-technical imaginary, envisioning is an intentional practice and process leading up to a social vision of a desirable future. By envisioning I therefore mean the process of discussing and agreeing upon a shared desirable future, including its associated values and purposes (ends). The act of envisioning is antithetical to hierarchically transmitted imaginaries stemming from the interest of corporations striving for capital accumulation and other values and purposes only in proxy of this interest. It is also antithetical to

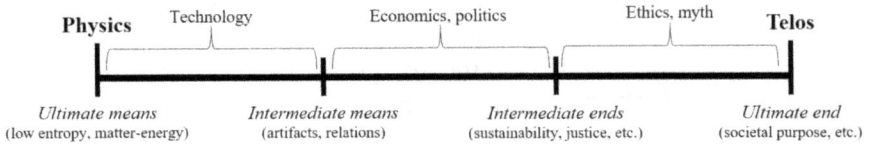

Figure 7.1 A depiction of the relation between physical reality and telos. Inspired by Daly and Farley's "means-ends spectrum" (2010: 49). The separation of the spheres technology, political economy, and ethics/myths does not mean that these are unrelated. As I argued in Chapter 2 and 3, a social metabolism requires not only technological artifacts for facilitating metabolic exchange with nature, but also social relations and cultural categories supporting and reproducing such metabolic exchange.

the theater of government politics void of discussions on the purpose of having a society. Understood this way, it is my conviction that envisioning is a powerful tool for aiding a wide social-ecological transformation rising bottom-up. For such envisioning to have teeth, however, it must be realistic and reflect the true transformative potential of the energy technology upon which it depends.[2]

The "means-ends spectrum" of ecological economists Herman Daly and Joshua Farley (2010: 48–50) is useful for better understanding realistic envisioning. Realistic envisioning essentially involves creating an alignment between the "ultimate means" and the "ultimate ends" (see Figure 7.1). It includes, in other words, figuring out what and how much matter-energy is necessary for achieving a social group's self-articulated purpose (or ultimate end). It also includes figuring out what societal purpose is suitable – i.e., realistic – given the physical possibilities and constrains of that social group's environment. In between the physically realistic conditions and the envisioning of the desirable ends are the "intermediate means and ends." These include the technological artifacts and infrastructure and the desirable values of a social group. Arguably, modern peoples are better equipped to have conversations about intermediate ends and means – such as solar PV modules and sustainability – than the societal purpose that they are meant to achieve or the matter-energy that they require. Ultimate means and ultimate ends are more or less taboo in modern societies, as exemplified by the almost complete absence of acknowledgment of limits to growth (ultimate means) and its implications for the tacit societal purpose of capital accumulation (ultimate end) in conventional politics.

Conventional visions which include solar PV technology often take the technology as a physically given. The obstacles to solar PV development are therefore considered to be of a political – not biophysical – nature. Despite much research attempting to raise awareness of the matter-energy constraints of solar PV technology (see Chapter 1), most of the conventional solar visions discussed in the previous chapters are based on the notion that nature provides solar power for free. Matter-energy constraints are rendered irrelevant for realizing a just and sustainable society based on energy harnessed through solar PV technology. As we have seen throughout this book, this has led to a dissonance between the vision and reality of solar PV development which developers and "green" entrepreneurs are prone to deny (see also Goldstein, 2018). To avoid such dissonance, I suggest the following (broadly formulated) steps to facilitate realistic envisioning:

1 Acknowledge that technological artifacts, such as PV modules, require matter-energy and that human societies are materially bound to nature (see Chapter 2). Human-environmental separation, such as "decoupling," is therefore an unrealistic end.
2 The second step includes an assessment of the biophysical requirements of such technological artifacts (intermediate means) to find out whether the social group can provide for the necessary resources within the carrying capacity of their environment.
3 The third step can be approached in two different ways:
   a If the answer to question two is yes, the third step involves envisioning the desirable values and social relations that the technology may facilitate. This step also includes articulating desirable relations with non-human animals and other species.
   b If the answer to question two is no, the third step involves assessing whether the exchange relations necessary for the technology are aligned with the social groups desired values and social relations. This also involves re-considering the social group as larger than initially assumed and include voices from this broader social group in the envisioning process (to ensure recognition and procedural justice, Jenkins et al., 2016).
4 The fourth step include a reflection on what ultimate end, e.g., societal purpose, the realistic values and social-ecological relations may lead to. For the vision to remain realistic, any attempt at reimagining the physical constituents of the technological artifact or the social relations and values it implicates is best avoided in this step.

The steps are presented in the order of "means to end," starting with an acknowledgment of a physical world and ending with considerations of ultimate ends. This reflects the assertions of philosophical materialism, which safeguards against the notion that human cognition stands above or apart from nature as a means to dominate it or cheat it (see Chapter 2). It also safeguards against the idea that any intermediate means, e.g., energy technology, is suitable within the environment of a given social group. One important exception to this "means to end" trend is the third step, which is best understood as a two-way conversation between intermediate ends (values, desirable outcomes) and the intermediate means (social relations, artifacts). This is where idealistic envisioning encounters physical realism and where the tensions between the two culminate.

A technological artifact should not be accepted simply because it appears to be physically realistic. A blind acceptance of technological artifacts is associated with technological determinism, which is the idea that technological development determines social change. Technological determinism implies that human history is subjugated to technological change in such a way that socially envisioned values and social relations cannot be intentionally realized. Technological determinism, in this view, is part of the modern worldview and narrative that must be challenged. From another critical viewpoint, it is possible to understand technological determinism as a rhetorical tool for legitimizing commodification of previously unexploited aspects of social life or the natural world. In contrast to the bleak and apolitical outlook of technological determinism, Ivan Illich (1973), Ernst Schumacher (1993[1973]), Andrew Feenberg (1991) and contemporary degrowth scholars (e.g., Kerschner et al., 2018) have argued that desirable values and social-ecological relations *can* determine the choice of technology. This is a powerful notion for empowering individuals, communities, and organizations to think of desirable social relations as including the possibility of certain technologies which they can choose to pursue or not.

Realistic envisioning is based on the idea that social relations embody the possibility of a range of technologies, just as much as technologies embody certain social-ecological relations. The choice of energy technology can therefore technically be envisioned either "means to ends," emphasizing the possible range of social relations embodied in a physically realistic energy technology, or envisioned "ends to means," emphasizing the possible range of energy technologies embodied in the desirable values and relations. The conventional recipe for human development is based solely on the former view, which knowingly or unknowingly sacrifice socially desirable values and social-ecological relations by accepting new energy technologies as inherently "good" (instrumentalism). Freeman Dyson's notion that technology is a gift from God is a forceful notion allowing any technology to determine social change and diminishing the significance of desirable human values. Some solar visionaries are willing to embrace this notion while denying that technologies are dependent on matter-energy (see Chapter 2). Rather than being rooted in technological determinism, the "means to ends" approach in realistic envisioning is rooted in a physical assessment of the technology, which functions as a basis for an intentional social evaluation.[3,4]

Realistic envisioning is a dialectical notion. This means that on one level, the means and ends for a metabolic transformation are the same thing, whereas on another level, the tensions between the two are at the foundation of social-ecological change. On the one hand, only realistic visions can be achieved in the physical world. Desirable visions that are not realistic are therefore pointless.[5] On the other hand, the outcomes of physical assessments are meaningless outside of socially articulated visions, which they are set to evaluate. But neither physical assessment nor socially desirable visions are sufficient for agreeing upon the future's energy technology and shaping the course of history. The two must conjoin in the world through the practice of real people. It is therefore of great importance that individuals, communities, and organizations test and practice a variety of energy technologies to be able to assess their transformative potential. It is only through actively engaging in the world – while being aware of the potentially global social relations implicated in the energy technology – that the full range of possible values and relations associated with the energy technology can be known. The truth-value of visions are ultimately decided historically, in a world that is constantly changing.

Alternative solar technology

This book demonstrates how conventional solar visions are contradictory because decision makers and developers are subjected to machine fetishism (see 'fetishism' in the Glossary). The physical evidence for this conclusion did not exist 20 years ago when these visions were formulated. The solar PV industry was then still in its infancy and solar visions for democratic and sustainable energy remained improbable and marginal. The path of finding a socially just and ecologically sustainable energy technology has to go through the formality of occurring (see Chapter 2). Having now occurred, the rapid development and installation of solar PV modules reveals that this technology was not hostile to the process of capital accumulation reproducing global environmental injustices. Solar PV technology, we now know, is not a local energy technology at all, but heavily contingent on global price differences on energy and materials (such as polysilicon and silver), globally cheap labor and government strategies for climate mitigation, energy independency and imperial-like strategizing. Could solar PV technology have developed in any other way? History could certainly have been different, but the results of this book imply that the biophysical prerequisites of solar PV technology meant that it was always better suited as a commodity for the world market than as a democratic and sustainable tool for local energy harnessing. "Made on the world market" was written on the components from the very start.

The question of whether the energy technology for a socially just and ecologically sustainable energy transformation can and should emerge from a world market is still a contested issue. For some, the global prerequisites of solar PV technology does not present a problem for a just and sustainable energy transformation, because the uneven distribution of work and resources is not considered an essential property of the world economy (see Chapter 3). From this perspective, international cooperation and trade must increase under reform as the world is heading toward greater unity. For others, a resource-heavy global transportation system exploiting global price differences was never part of the just and sustainable energy transformation that they envisioned. Such a system, moreover, is antithetical to increased unity, which must be based on trust cemented by equal exchange. This book shows that ecologically unequal exchange facilitates an uneven transfer of resources in the global solar PV industry in such a way that some people now enjoy the positive benefits of solar power while others bear the burdens. This condition is not likely to change (see Chapter 5). It may be too late and too big of a task to change the conventional solar visions rooted in the notion that solar power is inherently just and sustainable. Transforming the global system of exchange developed since the 16th century may also prove a futile task.[6] It appears realistic, however, to envision and practice alternative solutions in the active hope that another form of solar technology demonstrates better prospects of providing energy to local communities without compromising the values of social justice, equality and ecological sustainability.

The technological "artifacts" of this alternative solar technology are not mechanical objects assembled in world markets, but living plants assembled in local ecosystems. Similar to solar PV modules based on the photoelectric effect, plants based on photosynthesis have the capacity to harness direct sunshine for human use. Contrary to solar modules, however, plants do not necessitate a system of international trade (unless biogenetically synthesized or industrially grown). Their roots do not extend out in the world economy, but are firmly anchored in local soils and local human care. The history of agricultural practices in diverse ecosystems means that this alternative solar technology have already undergone the formality of occurring. In an effort to compare what Staffan Delin (1985) calls the "technology of nature" with the "technology of humans," he concluded that plants are

> approximately a billion times better than our industrial technology when it comes to putting atoms into useful structure. Therein lies the explanation to the fact that they can draw on sunlight in order to build themselves from waste. ... It is simply not possible to develop a better technology than that which they are already using. ... In turn, this means that we already now have access to the best technology we could possible imagine (Delin, 1985: 49–50).

The remarkable capacity of plants to draw on sunlight to build themselves from waste has been gradually improved over countless generations during the course of evolutionary history. There is also a long historical effort of improving the human relation with plants as living energy collectors, a practice as diverse and rich as human communities on Earth. Plants, as such, are the foundation of a radically alternative solar technology, reproduced through the energy technology of subsistence-oriented practices such as foraging, horticulture, or traditional and small-scale agriculture.

Having identified photosynthesizing plants as the foundation for alternative solar technology, it is crucial to distinguish between industrial agriculture and traditional and small-scale agriculture (henceforth, traditional agriculture). Industrial agriculture, contrary to traditional agriculture, is not an energy technology, but primarily a way of dissipating energy (Martinez-Alier, 2011). Whereas traditional agriculture developed as a means to *harness* energy, the development of industrial agriculture relied on highly energy-intense inputs, such as artificially manufactured fertilizers, which transformed agriculture from a means to "produce" energy into a means of "consuming" energy.[7] Thus, the EROI of traditional food systems are usually net-positive whereas industrial agricultures are substantially lower and often net-negative (Martinez-Alier, 2011; Smil, 2016). A net-negative energy return implies that industrial agriculture cannot reproduce the biophysical conditions that it requires. It is, in other words, dependent on a net import of energy and resources in order to produce food and materials. Industrially grown plants, much like solar PV modules, are heavily dependent on international trade for its necessary resources. Wild or traditionally grown plants are not restricted by this social condition.

Alternative solar technology utilizing the photosynthesis of plants has important physical limitations. First, like any other energy technology, it is limited in the range of realistic energy conversions. Burning agricultural produce to propel cars or tractors is one such unrealistic energy conversion (see Chapter 5). This is not to say that biomass of photosynthesizing plants cannot be used for a variety of ends, including human and animal metabolism (eating), heat and electricity generation (burning), and infrastructure (houses, boats, irrigation systems). Considering the limited uses for biomass, this alternative solar technology would need to be complemented with other ways of utilizing available local energy sources, including passive solar heating, conversions of wind energy (windmills, windpumps), utilization of water energy (waterwheels, water pumps), low-tech water batteries, and irrigation systems, to mention only a few. Second, despite its high energy efficiency, the land requirements of alternative solar technology remain high per watt capacity. According to Smil (2015: 80),

> The photosynthetic conversion of select wavelengths of solar radiation to the chemical energy of new phytomass [biomass] is a remarkable transformation, but one with inherently low efficiency [in terms of surface area] because only a small part of solar energy that is initially converted to new chemical bonds in those plant tissues ends up as harvestable phytomass.

Smil (2015: 87) estimates a power density of 0.6 W/m^2 for burning wood for heat. While this is near twice as efficient as the power density$_{ext}$ of solar PV technology (see Table 5.9), it implies that high levels of energy consumption cannot be sustained through alternative solar technology without some form of appropriation of other people's resources (a frequent phenomenon in history, see Chapter 3). This limitation dictates that alternative solar technology needs to be based on social aspirations similar to "degrowth," in which the reduction of energy throughput is aligned with desirable ends (see below). Solar PV technology could be utilized in a similar way, but would then, contrary to alternative solar technology, undermine the possibility for its own reproduction. Solar PV technology requires the extraction and processing of materials that are nonrenewable and exist only in limited quantities scattered throughout the world. Their continued extraction therefore requires a growth economy that can penetrate ever-more remote regions of the world. In contrast, the materials for alternative solar technology is geographically available in even the most inaccessible places on Earth and is therefore compatible with the ambition to sustain social groups with lower levels of metabolic throughput. Given these limitations, it is important to carefully think about the physically im/possibility of various (ultimate or intermediate) ends when envisioning the option of alternative solar technology.

Alternative solar technology is associated with many more socially and ecologically favorable possibilities. Managed with care for both humans and non-humans, traditional-type farming based on sustainable and regenerative practices can improve soil quality, sequester carbon, reduce surface temperatures, prevent erosion and enable a high diversity of flora and fauna in local ecosystems (Hathaway, 2016; Maezumi et al., 2018; Kerr et al., 2021). If peasants do not (exclusively) produce for an external market, traditional agriculture also has the potential to close the "metabolic rift" whereby nutrients of soils are not returned to the land (Foster et al., 2010). We should remember that the modern world division of labor became increasingly significant when Europeans appropriated fertilizers (guano) from their colonies as a response to exhaustion of nutrients in their own rural areas (Clark and Foster, 2009). Traditional agriculture has the potential to circulate nutrients in local ecosystems and may thereby function as a means to reverse this long historical trend. By relying on energy captured through traditional agriculture, people may therefore challenge the injustices generated by ecologically unequal exchange and the institutions of international trade by once again circulating nutrients in local ecosystems.

The rise of transnational agrarian movements is cause for taking alternative solar technology seriously. There are today more peasants in the world than any other time in human history (van der Ploeg, 2014). The most prominent of these movements, La Vía Campesina (i.e., The Peasant Way), includes more than 200 million peasants in 81 countries. Founded in 1993, the movement seeks to connect peasants, indigenous peoples, pastoralists, fishers, rural women and peasant youths throughout the world to fight against transnational corporations, free trade capitalism and patriarchy. The movement is simultaneously working for climate and environmental justice, peasant's rights, agrarian reform and "food sovereignty." Food sovereignty, the movement's primary demand, refers to

> The right of peoples to healthy and culturally appropriate food produced through ecologically sound and sustainable methods and their right to define their food and agriculture systems (La Vía Campesina, 2021).

Food sovereignty is a democratic demand, seeking for peasants to re-establish the right and autonomy to distribute the food that they produce and the energy that they harness. The movement envision their cultivation practices as distinct from the world market and is politically opposed to international regulatory organizations such as the World Trade Organization (WTO), who present obstacles to peasant sovereignty, regenerative (non-extractive) farming and democratic organization. The movement has gained prominence primarily in the Global South, but is rapidly spreading also in the Global North.

In terms of energy technology, the movement released a paper in 2007 explaining how "small scale sustainable farmers are cooling down the earth." By using farm practices that store CO_2 and use considerably less energy on farms, the movement implied that peasant farming may be a solid foundation for an alternative energy regime (La Vía Campesina, 2007). In the paper, the movement listed five points on how traditional farming is superior to industrial agriculture.

The fifth point noted how traditional cultivation practices have the capacity to "transform agriculture from an energy consumer into an energy producer" (ibid.). The paper also focused on how traditional agriculture makes possible a form of decentralized energy production and consumption. This demonstrates how the movement, having articulated clear visions for a socially just and sustainable society, is open to the means of traditional farming as an energy technology for realizing them.

Among numerous other transnational agrarian movements, the permaculture movement is gaining increased popularity worldwide. Similar to La Vía Campesina, the permaculture movement argues for agroecological practices in resistance to the ecological devastation of industrial agriculture. Contrary to La Vía Campesina, the permaculture movement is founded on a more unified set of principles for practicing healthy human-environmental relations in diverse ecosystems (see, e.g., Hathaway, 2016). It lacks the organizational structure and unifying emblems characteristic of La Vía Campesina. Instead, as a social movement, permaculture operates in "various groups in embedded 'fragmentary' networks 'submerged in everyday life'" (Leahy, 2021: 49). The movement is rooted in a combination of indigenous knowledge, Masanobu Fukuoka's "natural farming," and scientific perspectives on energy flows in ecosystems observed by Howard T. Odum (Fukuoka, 2009[1978]; Mollison, 1979). More importantly, however, it relies on continued observation and adaption to the interrelations of local ecosystems.

The permaculture literature shows a relatively high level of energy literacy. This is reflected in some of the more popular books of the movement (Mollison, 1979) and in studies on energy in permaculture systems. In one study on sustainable energy in South African permaculture, Elisabeth May Kruger (2015) concluded that permaculture systems have the potential to

> maximize passive energy flows, and broadens conceptions of 'energy', its use and storage, beyond provision of electricity and to include kinds of energy that are essential to autarky and self-sufficiency in the form of air and water, as well as other natural resources like food and building materials.

The notion of conceptualizing energy "beyond provisioning of electricity" is an important notion for considering cultivation as a form of energy technology. The finding that such energy technologies are compatible with "autarky and self-sufficiency" is an important finding demonstrating the local potential for plant cultivation as a democratic and sustainable form of energy technology. In Terry Leahy's newly published book *The Politics of Permaculture*, we also learn that permaculture practitioners often share the view that the ecological crisis "means that we have to move to a local agriculture and energy supply." While the notion of permaculture as a form of energy technology may not be directly obvious to practitioners, this demonstrates that the connection is certainly there. How extensive and how deep this connection goes, however, is an exciting subject for future study and engagement.

Together with other transnational agrarian movements,[8] La Vía Campesina and the permaculture movement represent the possibility of a material foundation for a metabolic counter-regime. Contrary to solar PV modules, the material resources necessary for traditional agriculture or other non-industrial modes of energy provisioning can be found within almost any ecosystem on Earth. The prospects for democratic energy provisioning is thereby higher, because traditional agriculture does not – like solar PV development – necessitate large amounts of matter-energy imported from other ecosystems and peoples. This physical dimension to alternative solar technology also grants it an anti-exploitative and anti-capitalist potential. In the words of van der Ploeg (2014: 119):

> Peasant agriculture can enter where capital cannot go … It goes to the altiplanos of Peru and Bolivia, steep slopes and wet areas elsewhere, the *bolanhas* of western Africa and *baldios* in the north of Portugal where the costs of cultivation would be far too high to provide even an average return on capital. Such areas are not attractive to capital. Large areas of the world fall into these categories.

Such an anti-capitalist potential runs contrary to local communities based on solar PV modules, whom, as we saw in the case of Katrineholm, are dependent on a direct connection to the global market for their matter-energy provisioning. This does of course not mean that traditional agriculture is by default independent from the international market. Many of today's traditional and small-scale agricultural farms are producing food for the global market. In a famous critique of "food sovereignty," Henry Bernstein (2016: 132) argued "that there are no 'peasants' in the world of contemporary capitalist globalisation." Crucially, however, this does not foreclose that traditional agriculture *can* operate independent from the contemporary capitalist market, nor that there *are* prominent social movements currently fighting for this intermediate end. This calls for deeper engagements – both scientific and practical – with the transformative potential of traditional agriculture as an energy technology with potential for an energy transformation with more sustainable and democratic outcomes.

The articulated ends of "degrowth" are naturally compatible with the means of alternative solar technology, yet few are working toward understanding or generating their alignment. Degrowth can be understood as a new form of social-ecological regime operating to fulfill the needs of humans and environments under a radically reduced metabolic throughput (see 'throughput' in the Glossary). As the name suggests, degrowth refers to an aggregate reduction in production and consumption of commodities. By extension, however, it also refers to generate a socially just and ecologically sustainable society in which relations are *different*. "The objective is not to make an elephant leaner, but to turn an elephant into a snail," degrowth proponents argue (Kallis et al., 2012). The transition to a degrowth society is envisioned to be based on more convivial relations supported by common ownership, work sharing, local food production, basic income, local currencies, low-tech and greater emphasis on happiness and care

(Demaria et al., 2013; Kallis et al., 2018; Mehta and Harcourt, 2021). Much research has been conducted on the appropriate technology for a degrowth society, yet very few (if any) have investigated the notion of traditional-type agriculture as a foundational energy technology for degrowth communities. While some studies suggest that advanced technological options, such as wind power and solar PV modules, are realistic and desirable options for degrowth communities (e.g., Kunze and Becker, 2015), others demonstrate that the contradictions of relying upon such advanced technologies from global markets are incompatible with the values of the degrowth movement (Tsagkari et al., 2021). In agreement with the conclusions of this book, the latter study showed that the investigated communities remained "highly dependent on … global capital and knowledge flows" (ibid. 8). As such, the question remains whether any other form of energy technology exists and whether it is compatible with socially articulated visions for degrowth. Traditional agriculture signifies a potential energy technology for achieving this, but this possibility warrants more research and practice.

Before concluding this sub-section on the social-ecological ramifications of alternative solar technology, I would like to touch upon some questions and critical perspectives. The fact remains that traditional agriculture once was proliferating, but that its role as an energy technology transformed with the transition to the industrial regime. Does this historical turn mean that communities based on agrarian metabolisms risk taking this turn once again? While access to fossil fuels may be a one-time historical event, there is a very real risk and possibility that communities relying on alternative energy technologies have their produce appropriated by actors seeking profit maximization through global market exchange. Looking at the history of non-industrial societies in the world system, it was the colonial forces of capitalism that ended the reign of agriculture as an energy technology. Studies on the ecologically unequal exchange have shown how agricultural produce, forests (wood), food and a range of other resources generated through traditional agriculture have systematically been exported from the peripheries to supply the development of core nations (Hornborg, 2006; Shandra et al., 2009; Infante-Amate and Krausmann, 2019; Tasmim et al., 2021). Anthropologists, moreover, have shown how the introduction of advertisement and cheap produce from the international market eroded the conditions for previously just and ecologically sustainable economies based on traditional knowledge and practice (Norbert-Hodge, 2000). There is no question, then, that there are global economic forces at play seeking to appropriate the resources and energy harnessed through alternative solar technology. This, as I argue below, is why a metabolic counter-regime must be based on strategic alliances between social movements.

Alliances for a metabolic counter-regime

In order to generate a just and ecologically sustainable future, it may be necessary to form a metabolic counter-regime. The notion of a metabolic counter-regime builds on Karl Polanyi's "double movement," which explains how modern society is characterized by a dynamic interplay between the socially eroding and ecologically destructive "movement" of unregulated free market capitalism and the "countermovement" constantly seeking to protect society and nature (and also, paradoxically, the "movement" itself) (Polanyi, 2001[1944]). According to Polanyi (ibid. 137), "the countermove consisted in checking the actions of the market in respect to the factors of production, labor, and land." Since then, social scientists have identified the dynamics of this double movement also in contemporary society (e.g. Hann, 2007; Carton, 2014). The fact that the reactionary "countermove," not the proactive "movement," represents the destructive force is revealing. This implies that the forces to protect and ensure social justice and environmental sustainability are in fact also contingent on the destructive movement of unregulated free market capitalism. It is not simply the "movement" that is contingent upon the "countermovement" to check its actions and regulate its mindless greed for commodification and exchange value, but also the "countermovement" which relies on the material provisioning that this exploitation generates in society. Many urban climate justice activists still eat industrially produced food (even if they are from the dumpster) and may even in some regions rely on electricity or heat generated from coal. The notion of a metabolic counter-regime envisions the countermovement with a strategy for its own material provisioning (e.g., alternative solar technology) that grants it with an independence from – and therefore a proactive edge in relation to – the social and environmental destruction of unregulated market actors.

Let us recall that metabolic regimes are specific metabolic patterns of interaction between human societies and natural systems operating with specific social relations and cultural categories. The dominant metabolic regime, the industrial regime, is defined by a set of problematic conditions symptomatic of the "movement" of unregulated capitalism. This includes the reproduction of a world division of labor through the mechanism of ecologically unequal exchange (see Chapter 3), which concentrates technological infrastructure in the core nations of the world economy. Having concluded that it is difficult to separate the metabolic reliance on solar PV technology from ecologically unequal exchange and global price differences, it appears unrealistic that solar PV modules, or any other advanced renewable energy technology, can form the basis of a new metabolic regime based on social justice and ecological sustainability. This is nevertheless where the world economy is most likely heading, while the inaction of world leaders and corporations legitimizes the madness of burning more fossil fuels. In this situation, it is not sufficient, or even possible, to generate a new metabolic regime merely by protecting nature by "checking the actions of the market." What is needed, I would argue, is an alternative metabolic basis that is already compatible with the values of social justice and ecological sustainability, which can provide the countermovement with the means to act *metabolically independent* from the capitalist market.

Agrarian movements, based on alternative solar technology, may be capable of providing the material foundation for a metabolic counter-regime (for a similar call, see Dale, 2019). This metabolic foundation would come in the form of material provisions such as nutritious foods, building materials, clothing and transportation, but also in the form of social provisions such as belonging, care, nature connectedness, individual autonomy, competence development, music and art. The material provisions are more essential for establishing freedom from market forces and the capitalist mode of reproduction, the condition upon which the social provisions can develop and flourish. But the social provisions, if we recall, are also necessary components for successfully reproducing social metabolisms and, more importantly, crucial for their social and political appeal. While agrarian movements may be capable of forming the metabolic foundation of a counter-regime, they are currently facing numerous obstacles that may only be overcome by building alliances across countermovements.

As a first step, this includes identifying what function each social -or environmental movement may serve in reproducing a metabolic counter-regime. For the sake of simplicity, it is possible to divide such functions into two categories: One type of function pertains to acts of working constructively for a new type of material provisioning and new laws. This includes new ways of growing food, new ways of building houses, new ways of reusing nonrenewable materials, new banking systems and currencies, new ways of transporting materials, new ways of legislating environmental destruction and protection, and so on. The second type of function pertains to acts of resisting and dismantling the industrial regime in areas of great material or symbolic importance. This includes resisting the expansion of fossil infrastructure, dismantling already existing fossil infrastructure, shutting down mines, organizing protests and civil disobedience against unjust and unsustainable legislations, boycott of industrial products, general or targeted strikes, and so on. Let us call these two types of movement functions "subsistence" and "resistance." While the "subsistence" function seeks to generate the social-material basis of a new regime, the "resistance" function seeks to protect this subsistence by inhibiting projects and legislations undermining its progress.

The importance for the emergence of a metabolic counter-regime is that social and environmental movements competent in specific functions work together in such a way that each movement provides the other with the resources that they cannot provide themselves. For example, agrarian movements may rely on urban activists to resist the expansion of extractive capitalism on their land, while in turn, urban activists rely on agrarian movements to provide them with food and refuge to hide or set up camp (see Table 7.1 for more examples of movements). By supporting each other this way, diverse social movements may come together to form a metabolic counter-regime working toward independence from the global market and the capitalist economy. As ecological economists have pointed out, such alliances must be based on "value pluralism" and represent a "pluriverse," through which each group acknowledge that there are multiple and sometimes contradictory values and human-environmental relations (Martinez-Alier, 2002; Kaul et al., 2022). This is arguably crucial for maintaining lasting alliances across movements working toward varied intermediate ends (values).

Considering the varied intermediate ends of contemporary countermovements, it is likely favorable if alliances for a new metabolism form under at least one shared ultimate end. The call for degrowth represents one ultimate end now

Table 7.1 A non-exhaustive overview of subsistence- and resistance-oriented movements suitable for a metabolic counter-regime alliance

Organization/ movement	Objectives	Practices	Provisions
Subsistence-oriented			
Transnational agrarian movements	Food sovereignty, environmental justice, women's rights, etc.	Peasant farming and organization, awareness raising	Agricultural produce, alternative solar technology, organizational platform
Permaculture movement	Social-ecological connectedness, sustainability, regeneration	Agroecology, agroforestry, sharing economy, sustainable design, etc.	Agricultural produce, alternative solar technology, connectedness
Fablabs, makerspaces	Free manufacturing	Reuse, recycle, redesign, upcycle	Accessibility to nonrenewable materials (already extracted)
Free software movement	Freedom for software users	Software development, awareness raising	Free and open-source software
Open-source hardware movement	Freedom for hardware users	Open-source labs, modular design	Open-source designs of hardware
Faculty for a future	Independent science for social justice and ecological sustainability	Connecting scientists, education, facilitating research	Education and research
Eco-villages, back-to-the-land, and – and transition town movement	Social-ecological sustainability, self-sufficiency	Subsistence agriculture, edible forests, awareness raising	Agricultural produce, housing solutions
Alternative banking and local currency initiatives	Financial independence, sustainable banking, access to seeds	Money lending, timebanking, seed banks, etc.	Funding, time provisioning, etc.
Freegan movement (incl. food saving, freeshops)	Subsistence, reduction of wastefulness	Dumpster diving, food sharing, guerilla gardening	Agricultural produce, food, clothing, materials
Libraries	Media and tool accessibility, knowledge preservation	Organizing and sharing media and tools, etc.	Media, tools, space for learning
Festivals	Entertainment, connectedness	Music and art, workshops	Art, music, connectedness, etc.

Organization/ movement	Objectives	Practices	Provisions
Resistance-oriented			
Climate and environmental justice movement	Climate and environmental justice, ecological sustainability	Mass protests, mass civil disobedience actions, awareness raising	Resisting fossil-industry, dismantling fossil infrastructure
Nature conservation movement	Protect and manage nature's species	Creating and maintaining nature reserves/ wilderness, lobbying, awareness raising	Legal advice, funding, resistance to industrialism
Radical environmentalism	Ecological sustainability, gender equality, animal and nature rights	Sabotage, protest, civil disobedience, direct action, awareness raising	Dismantling industrial infrastructure, wildlife protection
Feminist movement	Gender equality, reproductive rights, women's liberation, etc.	Protest, civil disobedience, direct action, awareness raising	Resisting and dismantling patriarchy
Decolonial movement	Epistemic liberation	Epistemic disobedience, critical analysis	Resisting hegemonic culture and worldview
Labor unions	Fair compensation, improving working conditions, etc.	Worker organization, negotiation, strike, awareness raising	Negotiation, organization
Squatters and occupiers	Right to the city, affordable housing	Occupy infrastructure and houses, protests, adverse possession	Housing, strategic action

rising in prominence among social movements. The importance, I would argue, is that such ultimate ends include a consideration of the material provisioning (ultimate means) through which the ends are to be achieved. The notion of alternative solar technology as a metabolic basis for a new regime is interesting, but requires further research and engagement. The degree to which alliances can be built on alternative solar technology may therefore need to be flexible in the short term. There may even be a certain wisdom in using current resources and dismantled infrastructure, wherever it is possible, to create a solid foundation for alternative solar technology to flourish in the mid- and long term (Giotitsas et al., 2022). In other words, it may be beneficial to generate what economist Amartya Sen (1959) called "landesque capital," i.e., lasting improvements to soils and the preconditions for farming, already today, for the purpose of aiding alternative solar technology for the future.[9] The point here is that strategies for a metabolic counter-regime may need to break down the path toward its ultimate ends into short-, mid-, and long-term goals. In the short or mid-term, solar PV modules may serve some use, even if they are arguably best thought of as temporary organs for achieving something more democratic, biotic, and lasting.

The strength of a metabolic counter-regime lies not only in that it generates community independence from the global capitalist market, but that it can serve as an alternative sphere of sustenance to capitalism, in which workers need to sell their labor to earn money to spend on goods and services necessary for their survival. The metabolic counter-regime, grounded in a democratic and sustainable form of material provisioning, supplies the countermovement (in its broadest sense) with something to offer people whom are currently trapped within the industrial regime. This includes, for example, low-skilled workers in the fossil-industry who often take such "dirty" jobs simply because they need money for sustaining their family within the market system. A successful metabolic counter-regime would be able to offer a different and likely more fulfilling mode of sustenance and family life based on organization around commons (see, e.g., Bollier and Helfrich, 2014; Kostakis et al., 2016). Such a metabolic counter-regime is both more meaningful and impactful than individual countermovements organizing only to "check the actions of the market" and, in the rare occasion, offer the possibility to represent some idealized worker interest.

The notion of alternative solar technology and a metabolic counter-regime may not be immediately attractive. A future society based on electricity harnessed through massive amounts of solar PV modules will arguably remain a more attractive vision for some time. However, if my assessment of solar PV technology is correct, conventional solar visions will at some point appear so implausible that a society based on alternative solar technology will start to appear both more realistic and attractive. Even if such a moment arrives (for some it already has), it is unreasonable to expect that modern peoples will conclude that solar PV modules are in fact not living up to their promises. Political attractiveness, after all, is not rooted in rational assessment, but firmly grounded in the realm of symbols, meaning making and existential desire generated by lived human conditions and social metabolism. The flourishing of some of the most unrealistic conspiracy theories during the COVID-19 pandemic is the definite sign of 21st-century human irrationality. Modern humans, we must conclude, are perfectly capable of believing in almost anything. On the one hand, this can be viewed as a tragedy in a world needing prompt rational solutions to the ecological crisis. On the other hand, it means that it is possible to improve the political appeal of any vision, including realistic ones.

Notes

1   This is an important distinction because, as we saw in Chapter 2, there is a deeply held belief saying that design is sufficient for technological development and progress. By extension, some blame the obstacles to the energy transition by way of solar power and other renewables on the lack of political will (e.g., Jacobsson and Delucchi, 2011).

2   All social groups require energy for their existence. Energy technology, understood as the means or strategy by which such groups access energy, is therefore not a choice but an inescapable part of any social group. To deny this, or to not give due weight to its significance, may lead to unrealistic visions of desirable futures.

3   Such assessments include the measure of the ecological footprint, carbon footprint, energy return on energy investment, power density, ecologically unequal exchange, material flow analysis, life cycle analysis, human appropriation of net primary production, and more.

4   After having done such physical assessment of solar PV development, I have shown that many socially desirable visions for solar power are contradictory to their envisioned purpose (i.e., unrealistic, as defined above). The purpose of the book does not include evaluation of solar PV development based on this physical assessment. This is something that is better left for individuals, households, communities, etc. to perform.

5   If they were not meant to be unrealistic, e.g., for entertainment, from the start.

6   Even if the attempt is commendable and the outcomes not pre-determined.

7   As we saw in Chapter 3, the energy technology of cultivation was replaced as the dominant energy strategy by the combustion engine and the extraction of fossil fuels during the third wave of fossil development (Table 7.1).

8   For more examples, see Borras et al.'s (2008) review of the origins and politics of transnational agrarian movements.

9   For an overview of such improvements, see Håkansson and Widgren (2014).

References

Bernstein, Henry. 2016. Food sovereignty via the 'peasant way': A sceptical view. In *Critical perspectives on food sovereignty: Global agrarian transformations*, edited by Marx Edelman, James C. Scott, Amita Baviskar, Saturnino M. Borras Jr., Deniz Kandiyoti, Eric Holt-Giménez, Tony Weis, and Wendy Wolford, vol. 2, 119–152, London, England and New York, NY: Routledge.

Bollier, David, and Silke Helfrich, eds. 2014. *The wealth of the commons: A world beyond market and state.* Amherst, MA: Levellers Press.

Borras, Saturnino M. Jr., Marc Edelman, and Chstóbal Kay. 2008. Transnational agrarian movements: Origins and politics, campaigns and impact. *Journal of Agrarian Change* 8(2–3): 169–204. https://doi.org/10.1111/j.1471-0366.2008.00167.x.

Carton, Wim. 2014. Environmental protection as market pathology? Carbon trading and the dialectics of the 'double movement'. *Environment and Planning D: Society and Space* 32(6): 1002–1018. https://doi.org/10.1068/d13038p.

Clark, Brett, and John B. Foster. 2009. Ecological imperialism and the global metabolic rift: Unequal exchange and the guano/nitrates trade. *International Journal of Comparative Sociology* 50(3–4): 311–334. https://doi.org/10.1177/0020715209105144.

Dale, Bryan. 2019. Alliances for Agroecology: From climate change to food system change. *Agroecology and Sustainable Food Systems* 44(5): 629–652. https://doi.org/10.1080/21683565.2019.1697787.

Daly, Herman E., and Joshua Farley. 2010. *Ecological economics: Principles and applications.* 2nd ed. Washington, DC: Island Press.

Delin, Staffan. 1985. *Naturens teknik och människans.* Stockholm, Sweden: LTs förlag.

Demaria, Federico, Francois Schneider, Filka Sekulova, and Joan Martinez-Alier. 2013. What is degrowth? From an activist slogan to social movement. *Environmental Values* 22(2): 191–215. https://doi.org/10.3197/096327113X13581561725194.

Feenberg, Andrew. 1991. *Critical theory of technology.* New York, NY: Oxford University Press.

Foster, John Bellamy, Brett Clark, and Richard York. 2010. *The ecological rift: capitalism's war on the earth*. New York: Monthly Review Press.

Fukuoka, Masanobu. 2009[1978]. *The one-straw revolution: An introduction to natural farming*. New York, NY: New York Review Books.

Giotitsas, Chris, Pedro H. J. Nardelli, Sam Williamson, Andreas Roos, Evangelos Pournaras, and Vasilis Kostakis. 2022. Energy governance as commons: Engineering alternative socio-technical configurations. *Energy Research and Social Science* 84: 102354. https://doi.org/10.1016/j.erss.2021.102354.

Goldstein, Jesse. 2018. Planetary improvement: Cleantech entrepreneurship and the contradictions of green capitalism. Cambridge, MA and London, England: The MIT Press.

Håkansson, Thomas N., and Mats Widgren, eds. 2014. *Landesque capital: The historical ecology of enduring landscape modifications*. London, England and New York, NY: Routledge.

Hann, Chris. 2007. A new double movement? Anthropological perspectives on property in the age of neoliberalism. *Socio-Economic Review* 5(2): 287–318. https://doi.org/10.1093/ser/mwl027.

Hathaway, Mark D. 2016. Agroecology and permaculture: Addressing key ecological problems by rethinking and redesigning agricultural systems. *Journal of Environmental Studies and Sciences* 6: 239–250. https://doi.org/10.1007/s13412-015-0254-8.

Hickman, Caroline, Elizabeth Marks, Panu Pihkala, Sustan Clayton, Eric R. Lewandowski, Elouise E. Mayall, Britt Wray, Catriona Mellor, and Lise van Susteren. 2021. Climate anxiety in children and young people and their beliefs about government responses to climate change. A global survey. *The Lancet Planetary Health* 5(12): e863–e873. https://doi.org/10.1016/S2542-5196(21)00278-3.

Hornborg, Alf. 2006. Footprints in the cotton fields: The Industrial Revolution as time-space appropriation and environmental load displacement. *Ecological Economics* 59(1): 74–81. https://doi.org/10.1016/j.ecolecon.2005.10.009.

Illich, Ivan. 1973. *Tools for conviviality*. Glasgow, Scotland: Collins.

Infante-Amate, Juan, and Fridolin Krausmann. 2019. Trade, ecologically unequal exchange and colonial legacy: The case of France and its former colonies (1962–2015). *Ecological Economics* 156: 98–109. https://doi.org/10.1016/j.ecolecon.2018.09.013.

Jacobsson, Mark Z., and Mark A. Delucchi. 2011. Providing all global energy with wind, water, and solar power: Part 1: Technologies, energy resources, quantities and areas of infrastructure, and materials. *Energy Policy* 39(3): 1154–1169. https://doi.org/10.1016/j.enpol.2010.11.040.

Jasanoff, Sheila, and Sang-Hyun Kim. 2009. Containing the atom: Sociotechnical imaginaries and nuclear power in the United States and South Korea. *Minerva* 47: 119–146. https://doi.org/10.1007/s11024-009-9124-4.

Jenkins, Kirsten, Darren McCauley, Raphael Heffron, Hannes Stephan, and Robert Rehner. 2016. Energy justice: A conceptual review. *Energy Research and Social Science* 11: 174–182. https://doi.org/10.1016/j.erss.2015.10.004.

Kallis, Giorgos, Federico Demaria, and Giacomo D'Alisa. 2012. Introduction: Degrowth. In *Degrowth: A vocabulary for a new era*, edited by Giacomo D'Alisa, Federico Demaria, and Giorgos Kallis, 1–18. New York, NY: Routledge.

Kallis, Giorgos, Vasilis Kostakis, Steffen Lange, Barbara Muraca, Susan Paulson, and Matthias Schmelzer. 2018. Research on degrowth. *Annual Review of Environment and Resources* 43(4): 1–26. https://doi.org/10.1146/annurev-environ-102017-025941.

Kaul, Shivani, Bengi Akbulut, Federico Demaria, and Julien-François Gerber. 2022. Alternatives to sustainable development: What can we learn from the pluriverse in practice? *Sustainability Science* 17: 1149–1158. https://doi.org/10.1007/s11625-022-01210-2.

Kerr, Rachel B., Sidney Madsen, Mortiz Stüber, Jeffrey Liebert, Stephanie Enloe, Noélie Borghino, Phoebe Parros, Daniel M. Mutyambai, Marie Prudhon, and Alexander Wezel. Can agroecology improve food security and nutrition? A review. *Global Food Security* 29: 100540. https://doi.org/10.1016/j.gfs.2021.100540.

Kerschner, Christian, Petra Wächter, Linda Nierling, and Melf-Hinrich Ehlers. 2018. Degrowth and technology: Towards feasible, viable, and appropriate convivial imaginaries. *Journal of Cleaner Production*. 197: 1619-1636. https://doi.org/10.1016/j.jclepro.2018.07.147.

Kostakis, Vasilis, Kostas Latoufis, Minas Liarokapis, and Michel Bauwens. 2016. The convergence of digital commons with local mnufacturing from a degrowth perspective: Two illustrative cases. *Journal of Cleaner Production* 197(2): 1684–1693. https://doi.org/10.1016/j.jclepro.2016.09.077.

Kruger, Elisabeth M. 2015. Options for sustainability in building and energy: A South African permaculture case study. *Energy Procedia* 83: 544–554. https://doi.org/10.1016/j.egypro.2015.12.174.

Kunze, Conrad, and Sören Becker. 2015. Collective ownership in renewable energy and opportunities for sustainable degrowth. *Sustainability Science* 10: 425–437. https://doi.org/10.1007/s11625-015-0301-0.

La Vía Campesina. 2007. Small scale sustainable farmers are cooling down the Earth. Published 9-11-2007. https://viacampesina.org/en/small-scale-sustainable-farmers-are-cooling-down-the-earth/. Accessed 11-07-2022.

La Vía Campesina. 2021. The international peasants' voice. https://viacampesina.org/en/international-peasants-voice/. Accessed 11-07-2022.

Leahy, Terry. 2021. *The politics of permaculture*. London, England: Pluto Press.

Leopold, Aldo. 1993[1953]. *Round river: From the journals of Aldo Leopold*. Oxford: Oxford University Press.

Maezumi, Yoshi, S., Diana Alves, Mark Robinson, Jonas G. de Souza, Carolina Levis, Robert L. Barnett, Edemar A. de Oliveira, Dunia Urrego, Denise Schaan, and José Iriarte. 2018. The legacy of 4,500 years of polyculture agroforestry in the eastern Amazon. *Nature Plants* 4: 540–547. https://doi.org/10.1038/s41477-018-0205-y.

Martinez-Alier, Joan. 2002. *The environmentalism of the poor: A study of ecological conflicts and valuation*. Cheltenham, England: Edward Elgar.

Martinez-Alier, Joan. 2011. The EROI of agriculture and its use by the Via Campesina. *The Journal of Peasant Studies* 38(1): 145–160. https://doi.org/10.1080/03066150.2010.538582.

Mehta, Lyla, and Wendy Harcourt. 202. Beyond limits and scarcity: Feminist and decolonial contributions to degrowth. *Political Geography* 89: 102411. https://doi.org/10.1016/j.polgeo.2021.102411.

Mollison, Bill. 1979. *Permaculture: A designer's manual*. Tyalgum, Australia: Tagari Publications.

Norbert-Hodge, Helena. 2000. *Ancient futures: Learning from Ladakh*. London, England: Rider Books.

Polanyi, Karl. 2001[1944]. *The great transformation: The political and economic origins of our time*. Boston, MA: Beacon Press.

Schumacher, Ernst F. 1993[1973]. *Small is beautiful: A study of economics as if people mattered*. London, England: Vintage.

Sen, Amartya K. 1959. The choice of agricultural techniques in underdeveloped countries. *Economic Development and Cultural Change* 7(3): 279–285.

Shandra, John M., Christopher Leckband, and Bruce London. 2009. Ecologically unequal exchange and deforestation: A cross-national analysis of forestry export flows. *Organization and Environment* 22(3): 293–310. https://doi.org/10.1177/1086026609343097.

Smil, Vaclav. 2015. *Power density: A key to understanding energy sources and uses.* Cambridge, MA: MIT Press.

Smil, Vaclav. 2016. *Energy and civilization: A history.* Cambridge, MA and London, England: The MIT Press.

Tasmim, Samia, Jamie M. Sommer, and John M. Shandra. 2021. Feed me! China, agriculture, ecologically unequal exchange, and forest loss in a cross-national perspective. *Energy Policy and Governance* 32(1): 29–42. https://doi.org/10.1002/eet.1959.

Tsagkari, Marula, Jordi Roca, and Giorgos Kallis. 2021. From local island energy to degrowth? Exploring democracy, self-sufficiency, and renewable energy production in Greece and Spain. *Energy Research and Social Science* 81: 102288. https://doi.org/10.1016/j.erss.2021.102288.

Van der Ploeg, Jan D. 2014. *Peasants and the art of farming: A Chayanovian manifesto.* Halifax, NS, Canada and Winnipeg, MB, Canada: Fernwood Publishing.

Wullenkord, Marlis. 2020. Climate change through the lens of self-determination theory: How considering basic psychological needs may bring environmental psychology forward. *Zeitschrift Umweltpsychologie* 24(2): 110–129.

Appendix A

International trade volumes and embodied resources in the German–Chinese exchange

Table A.1 Trade volumes and embodied resources in Chinese export of solar PV modules to Germany, 2002–2018

Year	Chinese solar PV module exports to Germany		Embodied resources			
	Export (kg)	Prices (USD)	Energy (GJ)	Land (ha)	Labor (h)	Emissions (tons CO_2-eq.)
2002	117,206	15,655,463	52,442	113	31,177	3,092
2003	129,521	20,093,089	57,953	124	34,453	3,417
2004	1,234,167	145,459,891	552,216	1,186	328,288	32,559
2005	3,963,096	458,772,870	1,773,248	3,809	1,054,184	104,550
2006	6,976,314	802,959,450	3,121,482	6,704	1,855,700	184,042
2007	31,539,586[a]	1,387,741,794	14,112,072	30,310	8,389,530	832,050
2008	65,828,307[a]	3,134,085,716	29,454,218	63,261	17,510,330	1,736,617
2009	158,658,712[a]	3,776,077,334	70,990,254	152,471	42,203,217	4,185,575
2010	375,570,340	7,663,736,157	168,045,193	360,923	99,901,710	9,907,921
2011	327,077,931	5,704,730,963	146,347,749	314,322	87,002,730	8,628,643
2012	211,272,309	2,131,681,388	94,531,682	203,033	56,198,434	5,573,575
2013	59,025,716	592,278,908	26,410,466	56,724	15,700,840	1,557,157
2014	58,083,459[a]	345,596,582	25,988,863	55,818	15,450,200	1,532,300
2015	75,283,694[a]	313,933,004	33,684,936	72,348	20,025,463	1,986,059
2016	111,149,346[a]	396,803,166	49,732,663	106,815	29,565,726	2,932,231
2017	100,319,671[a]	358,284,551	44,887,081	96,438	26,692,198	2,646,528
2018	155,734,982[a]	463,406,970	69,682,131	149,710	41,436,627	4,108,437

Sources: Data from Comtrade (2021), commodity codes 854140 and 854150. Embodied resources, including indirect land requirements, calculated by the author.

a Data on export (kg) missing from Comtrade (2021). Calculated based on prices (USD) from Comtrade (2021) and $/W from Nemet (2019: 156–157). A price of $0.3/W in 2017 was deduced from price trends displayed in Nemet (2019: 156), and a price of $0.25/W was considered for the year 2018 (Nemet, 2019: 156; Haegel et al., 2019).

Table A.2 Trade volumes and embodied resources in German export of solar PV manufacturing machine to China 2002–2018

Year	German solar PV machine exports to China		Embodied resources			
	Export (kg)	Prices (USD)	Energy (GJ)	Land (ha)	Labor (h)	Emissions (tons CO$_2$-eq.)
2002	106	38,000	57	0.03	36	3
2003	2,500	909,000	1,350	1	849	78
2004	7,300	3,843,000	3,944	3	2,479	229
2005	10,900	3,321,000	5,889	4	3,702	342
2006	28,100	12,458,000	15,182	10	9,543	881
2007	853,900	130,479,000	461,354	297	289,984	26,758
2008	1,9849,44	300,227,011	1,072,445	689	674,087	62,202
2009	2,448,306	312,712,046	1,322,795	850	831,445	76,722
2010	8,884,239	1,187,604,272	4,800,065	3,085	3,017,087	278,404
2011	11,555,939	1,944,456,603	6,243,558	4,013	3,924,397	362,126
2012	4,453,060	391,886,514	2,405,944	1,546	1,512,259	139,545
2013	2,214,644	277,799,826	1,196,550	769	752,093	69,400
2014	2,934,083	322,045,392	1,585,256	1,019	996,415	91,945
2015	2,379,973	285,157,862	1,285,876	826	808,239	74,581
2016	5,321,773	418,854,750	2,875,301	1,848	1,807,274	166,767
2017	5,892,225	613,486,356	3,183,510	2,046	2,001,000	184,643
2018	5,009,753	606,865,365	2,706,719	1,740	1,701,312	156,990

Sources: Data from Comtrade (2021) commodity code 8486 and Trend economy (2020) commodity code 903082. Embodied resources, including indirect land requirements, calculated by the author.

Appendix B
Further considerations for calculating "power density extended"

The requisite labor and capital costs for solar development ultimately also represent land requirements. As both labor and capital have incontrovertible spatial correlates in the land areas required to reproduce and generate them, a calculation of the land requirements of a given technology would be incomplete without including the spatial demands corresponding to the inputs of labor and capital in the technology's construction and operation. This corresponds to assumptions regarding the technological boundary made in the study of $EROI_{ext}$ (Prieto and Hall, 2013; Ferroni and Hopkirk, 2016; de Castro and Capellán-Pérez, 2020). In line with these approaches, an assessment of the conditions for a renewable energy transition must attend to the total spatial demands of the labor and capital required to construct and maintain a massive technological infrastructure capable of replacing fossil fuel technologies. To build energy technologies is not simply a matter of applying engineering knowledge to certain physical forces of nature, but of accumulating a material infrastructure for harnessing such forces. The human labor and raw material for this infrastructure are spatially dispersed in a global political economy of social-ecological exchanges that are notoriously difficult to trace.

To indicate the magnitude of demands on surface area, the person-years of labor time needed for a country's PV development can be multiplied with the average ecological footprints of the workers inside and outside that nation's borders. The required labor time could then be translated into hectares. While these ecological footprints would exist regardless of whether the workers produced and maintained solar panels, a country's planned PV capacity – i.e., the solar panels themselves – cannot exist without a supply of labor time representing eco-productive space under the given historical circumstances. The labor demand for solar PV manufacturing, installation, and servicing ranges from 18,400 (IEA, 2017; IRENA, 2017) to 37,286 (DOE, 2017) to 53,028 labor hours per MW. Based on a somewhat conservative average of these figures, I will assume an estimate of 35,000 labor hours per MW throughout the entire global commodity chain. These labor hours are dispersed in the world economy according to different tasks throughout the commodity chain (Table B.1). The ecological footprints of the aggregate labor hours vary by country depending on what labor tasks in the commodity chain are conducted domestically and what tasks are outsourced to other nations.

Table B.1 Employment distribution in select events throughout the global commodity chain of a solar PV module

Select labor tasks in the PV module commodity chain[a]	Percentage of labor hours (%)	Geography
Projects/studies	1	Domestic/international
Silicon processing	3	Domestic/international
Cell manufacturing	8	Domestic/international
Module assembly	30	Domestic/international
Solar tracker	22	Domestic/international
Components and inverters	9	Domestic/international
Installation	21	Domestic
Operation	6	Domestic
Total	100	—

Based on Llera et al. (2013: 266).
Llera et al. (2013) do not account for the labor necessary for the raw material extraction.

How do we calculate the spatial correlates of the extraction, transport, and manufacturing of the component parts of solar PV modules? The main obstacle to doing this is that economic exchange is conventionally measured in money. However, this complication can be overcome if we recognize that energy consumption and real GDP are causally linked (Warr and Ayres, 2010; Ayres, 2016; Sultan and Alkhateeb, 2019). Based on this observation, I use a method that translates money into energy and then translates energy into land (see also Prieto and Hall, 2013; Hornborg et al., 2019). Let us look at each translation in turn:

a   *Money to energy:* As the GDP of a country is proportional to its energy consumption, the money to energy ratio is straightforward (Ayres, 2016: 382–386). In physics, all production processes are transformations of materials, which require energy. Real GDP is a measure of a given economy's total production output (be it in goods or services). Since all production processes are transformations of materials that require energy, energy consumption is inextricably connected to GDP. As Hagens (2020: 6, footnote 2) put it, "Money is a claim on energy, materials, and many other things. But every single good and service which generates GDP requires some energy conversion." In short, energy is a physical production factor without which economic activity would simply not be possible.

b   *Energy to land:* The best available estimate of how much land is embodied in energy is the carbon footprint. This has been calculated by Wackernagel and Monfreda (2004), who estimate that the carbon footprint of fossil fuels is 1,050–1,900 ha/MW. This means that for every MW worth of fossil fuel infrastructure, 1,050–1,900 hectares of land is required. I will use the median figure 1,475 ha/MW. Importantly, this figure includes both the land for the physical infrastructure and the land for sequestering the carbon emissions associated with the output of that infrastructure. Today, 84% of industrial energy is generated from fossil fuels, which means that I will calculate 84% of the capital investment as having a spatial correlate (BP, 2020).[1] This method of calculating the footprint of fossil fuels includes the surface areas needed for carbon sequestration through reforestation, which make up most of the footprint (99.998%). These surface areas may not strictly speaking be necessary for the operation of large-scale solar PV projects, but they are nevertheless required for their long-term sustainability.

China: Calculation of the direct and indirect land requirements of ERI's (2015) renewable energy scenario 2030

Calculating the direct land requirements of China's plan to massively increase its solar capacity by installing an additional 1,000 GW, we can begin by calculating the power density$_{ag}$ with the coefficient above (5 W/m^2). This yields the figure 20,000,000 ha, representing the surface areas occupied by the solar PV modules, as well as the spacing between them, access roads and empty land within the park fencing (Table B.2).

As for labor, we can begin by noting how the increased automatization in solar PV manufacturing has led to a decrease in labor hours over the last two decades. This explains the relatively small percentage of labor hours (8%) designated to the task of "cell manufacturing" in the commodity chain of solar PV modules (Table B.1). This reduction, however, should not be confused with the increasing trend to outsource the manufacturing facilities to peripheral nations such as Vietnam and Indonesia (see Chapter 4). That said, Chinese corporations, such as GCL System Integration, are also investing heavily in domestic manufacturing capacity, boasting plans to manufacture the largest solar module factory in the world that will be located in the eastern Chinese province of Anhui (Bellini, 2020). Given

Table B.2 Actual and projected annual electricity output per energy source in China's national electricity system under a "high renewable energy penetration scenario"

Energy technology	2015 (MW)	2015 (%)	2030 (MW)	2030 (%)
Coal	822,495	54	1,052,150	26
Oil	1,166	<1	1,012	<1
Fossil gas	94,273	6	130,119	3
Hydropower[a]	305,589	20	522,063	13
Wind	114,880	7.5	1,103,944	28
Solar	42,025	3	1,048,858	26
Of which is PV	(41,731)	—	(1,041,516)	—
Nuclear	42,790	3	66,000	2
Biomass	77,838	5	18,794	<1
Other[b]	21,232	1.5	58,203	1
Total	1,522,288	100	4,001,143	100

Source: ERI (2015).
a Including "pumped hydro storage."
b Including geothermal, ocean, waste, biogas, and chemical energy storage.

this situation, it is hard to foresee how much of the anticipated 1,000 GW will be manufactured domestically contra internationally. Based on this qualitative assessment, I will assume that 85% of the solar PV modules emerge within a completely domestic commodity chain, while the remaining 15% are manufactured and assembled in Vietnam and Indonesia.

The total 1,000 GW multiplied by the total labor hours throughout the commodity chain (see above) amounts to 35,000 million labor hours. Assuming a 2,000-hour-long work-year, these labor hours represent 17,500,000 person-year equivalents, 85% of which are located in China and 15% of which are in part (38%) located in Vietnam/Indonesia. With an average ecological footprint in China of 3.6 ha and the average ecological footprint in Vietnam (2.1) and Indonesia (1.7) of 1.9 ha, this amounts to a requirement of 59,697,500 ha associated with Chinese workers and 1,745,625 ha associated with Vietnamese and Indonesian workers. Measured this way, this gives a total land requirement of 61,443,125 ha.

As for the capital investment, the International Finance Corporation estimates that the cost of solar parks is approximately 1.74 million US dollars per MW (IFC, 2015). IFC's benchmarks include labor costs but exclude costs related to FITs, interest payments, insurance, construction of access roads, administration costs, the construction of a potential backup storage system, and repairs following events such as floods, hurricanes, and wildfires. Subtracting the labor costs (120,000 per MW) and adding 0.5 million US dollars per MW for some of the above-mentioned expenses (see Hornborg et al., 2019; cf. Prieto and Hall, 2013; Ferroni and Hopkirk, 2016) yields an estimate of 2.12 million dollars per MW. Using this coefficient, the Chinese ambition to install an additional 1,000 GW corresponds to 2.12 trillion US dollars.

Given that 1,823,200 million US dollars (i.e., 84%) of capital investments in PV represent around 13.8% of China's GDP[2] and that China's total use of fossil energy is around 23,301 TWh per year (2,003,512 ktoe; IEA, 2020a), the country's investment in PV can be taken to represent the use of approximately 3,728 TWh of fossil energy. Dividing this fossil energy by 8,766 hours in a year and converting the quotient into MW, we get 366,747 MW. Considering a footprint of 1,475 ha/MW, this figure corresponds to 540,951,825 ha.

US: Calculation of the direct and indirect land requirements and power density$_{ext}$ of DOE's (2021) "Decarb+E" scenario by 2035

Turning to the land requirements of US's ambition to increase its solar PV capacity, we can begin by calculating the direct land requirements according to the power density$_{ag}$ above (5 W/m^2). Applying this figure to the 1,600 GW solar PV infrastructure equals an even 32,000,000 ha. This designates the surface area necessary for the solar PV parks, including the solar PV modules, spacing between them, and access roads.

As for labor, US will likely rely on imported solar PV modules from China and Chinese companies operating from Southeast Asia. Currently, the US demand is low enough for US manufacturers to provide one third of the country's demand for solar PV modules (DOE, 2022). Given the current administrations current policies, I will assume that 50% (800 GW) of US's planned solar PV capacity be manufactured domestically and that the remaining 50% (800 GW) will be manufactured in China or Southeast Asia (albeit installed and operated in the US). The US 2050 target will require approximately 28 million labor-years, 14 million of which will be work in the US and 14 million being labor embodied in imported materials and components from Asia. Multiplied with the ecological footprint of the US (8,1 ha per capita) and Malaysia (2 ha per capita), respectively, these figures represent a land requirement equivalent to 113.4 million ha (US) and 28 million ha (Asia), respectively. The land embodied in the required labor time thereby amounts to 141.4 million ha in total.

Concerning capital investment, the US's plan to install 1,600 GW of solar PV capacity translates into 3.39 billion US dollars (2.12 m$/MW). Given that 2.78 billion US dollars (i.e., 84%) of the US's capital investment in PV technology represent around 0.012% of the US's GDP[3] and that the US's total fossil energy supply is around 20,000 TWh per year (Ritchie and Roser, 2020) the country's investment can be taken to represent the use of 2,400 TWh of fossil energy. Dividing this fossil energy by 8,766 hours and converting it from energy consumption to energy capacity (MW), we get 273,000 MW. Considering the footprint of fossil energy (1,475 ha/MW), this corresponds to 402,675,000 ha.

Germany: Calculation of the direct and indirect land requirements in Germany's target scenario "KA65"

Turning to the land requirements of Germany's ambition to double its solar PV capacity, we can begin by calculating the direct land requirements according to the power density$_{ag}$ above (5 W/m^2). Applying this figure to the 50 GW solar PV infrastructure equals an even 1,000,000 ha. This designates the surface area necessary for the solar PV parks, including the solar PV modules, spacing between them, and access roads (Table B.3).

Apart from the direct land requirements, there are also the indirect land requirements of labor and capital. Let us look at labor first. As we saw in the previous chapter, Germany's rapid installation of solar PV modules hinged in large part on the access to cheap Chinese raw materials, fossil energy, and labor. Still in 2020, Germany is dependent upon this relation as 80% of all modules installed are imported to Germany from Asia (notably China) (Fraunhofer, 2020). These 80%, however, are both installed and operated in Germany. Assuming that the remaining 20% of the modules are manufactured in Germany, albeit with Chinese polysilicon, the German aspiration to install an additional 50 GW of solar PV capacity until 2030 will require approximately 350,000,000 labor hours, 58% of which will be work in low-wage nations like China and 42% of which will be work done in Germany.[4] Related to the ecological footprint of an average Chinese and German worker, 203,000,000 labor hours from China represents 197,622 hectares and 147,000,000 labor hours from Germany represents 533,556 hectares.[5]

Table B.3 Projected annual electricity output per energy source in Germany's national electricity system under the scenario "coal phase-out + 65% target (KA65)"

Energy technology	2017 (MW)	2017 (%)	2030 (MW)	2030 (%)
Coal	46,000	21	16,000	6
Oil	2,000	1	6,000	2
Fossil gas	30,000	14	33,000	12
Nuclear	10,000	5	0	0
Wind	57,000	27	98,000	37
Solar PV	43,000	20	93,000	35
Hydro	10,000	5	10,000	4
Biomass	8,000	4	5,000	2
Other	7,000	3	5,000	2
Total	213,000	100	266,000	100

Source: AER (2018).

This amounts to a total of 898,956 hectares for Germany's target to install 50 GW by 2030.

Calculating the land requirement of the capital investment, I apply the same method of calculation as described above in the case of China. This involves using the IFC (2015) benchmarks on the costs of utility-scale solar parks, albeit with a subtracted cost of the labor and the added costs of processes not included by IFC. The estimate reached above was 2.12 million US dollars per MW. In the case of Germany's 2030 aspiration to install 50 GW solar PV capacity, this corresponds to 106,000 million US dollars. The portion of this capital that derives from fossil energy (84%) is 89,040 million US dollars, which corresponds to 2.4% of Germany's GDP at the time of the proposal.[6] In the same year, Germany's total fossil energy use was roughly 2,904 TWh (IEA, 2020b). The country's investment in PV can thereby be taken to represent the use of approximately 69,696,000 MWh of fossil energy. Converted into MW, we get 7,951 MW, which corresponds to a footprint of 11,727,725 ha.

India: Calculation of the direct and indirect land requirements for India's 2030 target

We can begin by applying the power density$_{ag}$ coefficient (5 W/m^2) to the additional 184.3 GW solar PV capacity now planned in India. This amounts to 3,686,000 ha necessary for the direct infrastructure, including spacing and other open surfaces within the park limits (Table B.4).

As for the indirect land requirements of labor, despite efforts on behalf of the Indian government to favor domestic manufacturing of solar PV modules, most of the solar PV modules installed in India are imported from China (Gupta, 2020). As much as 85% of the solar PV modules are imported from China, Vietnam, or Thailand (Gupta, 2019; MERCOM, 2020). Given these developments and ambitions, I will assume that India will be able to reach a domestic production of 30%

Table B.4 Actual and projected annual electricity output per energy source in India's national electricity system under the "High Renewable Energy Scenario"

Energy technology	2019–2020 (MW)	2019–2020 (%)	2030 (MW)	2030 (%)
Coal, oil, gas	231,000	62	263,000	33
Hydropower	51,000	14	84,000	11
Wind	38,000	10	169,000	22
Solar	35,000	9	229,000	29
Of which is PV[a]	33,250	(9)	217,550	(28)
Nuclear	7,000	2	17,000	2
Biomass, waste	10,000	3	23,000	3
Total	372,000	100	785,000	100

Source: Spencer et al. (2020). Figures from the "High Renewable Energy Scenario."
a   Assuming 95% of the solar power capacity from PV (see IRENA, 2017: 56).

of the annual installed capacity during the next 10 years. The remaining 70% will be calculated as if they are imported from China, where labor tasks related to research, silicon processing, cell manufacturing, module assembly, manufacturing of solar tracker, and various electrical components are located (63% of all labor in commodity chain). Even if the solar PV modules are imported from China, they are installed and operated with Indian labor (27% of labor in commodity chain). The total 184.3 GW multiplied by the total labor hours throughout the commodity chain amounts to 3,225,250 person-year equivalents. Out of these labor-years, 56% will be located in India and 44% will be located in China. When multiplied with the ecological footprint per capita of the two countries, this amounts to 2,163,498 ha for the Indian labor and 5,120,406 for the Chinese labor. In sum, the indirect land requirement of the labor for India's solar PV aspiration amounts to 7,283,904 ha.

Using the coefficient of 2.12 million dollars per MW, India's capital investment in solar PV modules amounts to a sum of 390,716 million US dollars. The proportion of this money that can be said to derive from fossil energy (84%) is 328,201 million US dollars, which corresponds to 11.4% of India's GDP.[7] In the same year, India's total fossil energy supply was 7,762 TWh (667,385 ktoe; IEA, 2020c). India's investment in solar PV development can thereby be taken to represent the combustion of approximately 885 TWh of fossil energy. Converted into MW, this amounts to 100,958 MW with a footprint of 148,913,050 ha.

Italy: Calculation of the direct and indirect land requirements for Italy's 2030 target

Taking the power density$_{ag}$ of 5 W/m^2 as an estimate, the additional 31,438 MW PV capacity needed to generate 30% of the annual electricity in Italy would require 628,760 ha of land directly (Table B.5). As for labor, during the boom of the global solar market, Italy relied to a high degree upon imported solar PV modules from China. During this time, around 85% of Italy's demand for solar PV

Table B.5 Actual and projected annual renewable electricity capacity in Italy's national electricity system under the "Integrated National Energy and Climate Plan"

Energy technology	2017 (MW)	2017 (%)	2030 (MW)	2030 (%)[a]
Hydropower	18,863	12	19,200	11
Geothermal	813	0	950	0
Wind	9,766	6	19,300	12
Solar	19,682	13	52,000	30
Of which is PV	(19,682)	(13)	(51,120)	(30)
Bioenergy	4,135	3	3,760	2
Total:	53,259	34	95,210	55

Source: NECP (2019: 69–70). Does not include fossil technology capacity (coal, oil, fossil gas, etc.), hence the "missing" percentages.

a   Pertains to the estimated electricity generated (not the generation capacity).

modules was met through international trade (Terzini et al., 2011; Cai et al., 2017). There is little to suggest that this situation will change in the near future. I will therefore assume that 15% of Italy's planned solar PV capacity will be manufactured domestically and that the remaining 85% will be manufactured in China (albeit installed and operated in Italy). Italy's 2030 target will require approximately 550,165 labor-years, 208,791 of which will be work in Italy and 341,374 being labor embodied in imported materials and components from China. These figures represent a land requirement equivalent to 918,681 ha (Italy) and 1,228,946 ha (China) respectively. The land embodied in the required labor time thereby amounts to 2,147,627 ha in total.

Concerning capital investment, Italy's plan to install 31,438 MW of solar PV capacity translates into 66,649 million US dollars (2.12 m$/MW). Given that 55,985 million US dollars (i.e., 84%) of Italy's capital investment in PV technology represent around 2.8% of Italy's GDP[8] and that Italy's total fossil energy supply is around 1,429 TWh per year (122,894 ktoe; IEA, 2020d), the country's investment can be taken to represent the use of 40 TWh of fossil energy. Dividing this fossil energy by 8,766 hours and converting it from energy consumption to energy capacity (MW), we get 4,563 MW. Considering the footprint of fossil energy (1,475 ha/MW), this corresponds to 6,730,426 ha.

Notes

1   Since other energy sources have spatial correlates too, this means that the full footprint of the capital investment in PV is not calculated.
2   China's GDP was 11,064.67 billion US dollars in 2015 (Worldometer, 2020a).
3   US's GDP was 23,000 billion US dollars in 2021.
4   Assuming that an equivalent of 8,000 MW solar PV capacity is manufactured in China, but that 28% of the associated labor-hours are performed in Germany (research, installation, operation) and that the remaining 2,000 MW capacity is produced in Germany with 3% of associated labor-hours performed in China (silicon production).
5   The average ecological footprints of the German and Chinese populations in 2016 were 4.9 and 3.6 global hectares per person and year (GFN, 2019). I have calculated

a work year of 1,300 labor-hours in Germany in 2016 and a 2,000-hour work-year for Chinese laborers (OECD, 2019).

6   Germany's GDP (nominal) in 2017 was 3,693.2 billion US dollars (Worldometer, 2020b).

7   India's GDP (nominal) was 2,875.14 billion US dollars in 2019 (Trading Economics, 2020).

8   Italy's GDP was 1994 billion US dollars in 2017 (Worldometer, 2020c).

Glossary

Aspiration The articulated and pursued target of a specific social group. I use the term mostly to denote solar photovoltaic (PV) aspirations, i.e., the socially articulated targets to install solar PV technology.

Asymmetry Denotes the condition in which two categories are not the same, i.e., an imbalance of proportion. I use the word primarily to describe "global asymmetries," i.e., the uneven distribution of matter-energy determined by differences in wages and prices in the *world economy*.

Complexity Following Tainter (1988: 23), I use the term complexity to refer to "the size of a society, the number and distinctiveness of its parts, the variety of specialized social roles that it incorporates ... and the variety of mechanisms for organizing these into a coherent functioning whole." For Tainter (1988: 37–38), an increase in complexity, understood as an increase in "different kinds of parts, more social differentiation, more inequality, and more kinds of centralization and control," is a type of social response to problems. A substantial decline in the level of complexity is a social collapse.

Conception An abstract idea or categorization constructed by a social group to survive in – and/or to make sense of – the world. Following Bateson (2000 [1972]), human conceptions are continually developed in relation to the contexts (i.e., the natural environments and social networks) in which they are applied.

Dissipative structure A dissipative structure is a type of system – e.g., an organism, a hurricane, or an economy – which exists far from thermodynamic equilibrium. It maintains this state by drawing highly ordered (low-entropy) matter-energy from its environment and dissipating it in a manner that reproduces its form. A dissipative structure is an integral part of its environment since it depends upon it for this necessary exchange of *matter-energy*.

Ecologically unequal exchange The theory of ecologically unequal exchange explains how wealthier nations rely on net imports of resources to sustain their levels of consumption and technological development, while displacing much of their work and environmental loads to poorer nations. Ecologically unequal exchange occurs in international trade when prices of commodities are not proportional to the inputs of resources in their production. In such a scenario, commodities traded at an equal exchange value (say $100 for $100) may imply an unequal exchange in terms of biophysical measures – such as embodied land, labor, energy, or materials – spent in the production of the commodities.

Embodied   The biophysical resources dissipated (or, "invested") in the production of a given commodity. Embodied emissions, for instance, denotes the amount of greenhouse gas emissions (e.g., kg of CO_2-eq.) associated with the production of a given commodity (e.g., a solar PV module). The term "embodied land" denotes the land necessary for the production of a particular commodity. Throughout this book, I use the terms "embodied labor," "embodied energy," and "embodied resources" in a similar way.

Energy carrier   A substance or device that "contains" or "carries" energy potential for human end transformation. These include petroleum, batteries, electricity, mechanical springs, dammed water, ethanol, wood, and so on. Energy carriers are distinguished from "primary energy sources," which are the unprocessed stocks and flows of energy in nature, including direct sunshine, coal, crude oil, wind, and biomass.

Energy density   A measure of the amount of energy per unit of mass (e.g., measured as MJ/kg).

Entropy   According to the second law of thermodynamics, energy always tends toward a state of thermal equilibrium, i.e., toward less ordered states. For example, if a cup of hot tea is left alone, the tea will cool off until it reaches the temperature of its surrounding environment. Entropy is the physical tendency whereby highly ordered states (e.g., a hot liquid) naturally tend toward less ordered states (e.g., cold liquid). When energy is converted (e.g., from coal to electricity), there is always a loss of useful energy, which typically dissipates from the system in the form of heat. To regather the dissipated energy is theoretically possible, but the act of doing so requires more energy than is gathered and is therefore energetically futile. This means that no *dissipative structure* can sustain itself from the same energy indefinitely and is therefore inescapably dependent upon its environment for a continual supply of highly ordered (low-entropy) energy. Georgescu-Roegen (1975) famously suggested that not only energy but also matter is subjected to entropy.

Environmental load displacement   A situation wherein typically wealthier nations displace the environmental burdens resulting from their high levels of consumption and technological development to poorer nations in the world economy. Environmental load displacement generally occurs through *ecologically unequal exchange*, but it can also occur through waste dumping.

Exergy   According to the first law of thermodynamics, energy cannot be created or destroyed. It is therefore incorrect to talk of "energy consumption." What is consumed, strictly speaking, is "exergy," which refers to a thermodynamic system's total amount of usable energy. For readability, I have chosen to use the term "energy consumption" in this thesis, even if I actually mean "exergy consumption."

Fetishism   A cultural attribution of agency or inherent powers to inanimate objects. The word comes from the Portuguese idiom feitiço, meaning "spell" or "charm." Historically, Portuguese merchants used the word to describe religious practices among peoples with whom they traded along the west coast of Africa in the 15th century. Karl Marx later applied the concept to the modern understanding of commodities under capitalist relations of production (Marx, 1990[1867]: 163–177). Marx argued that the human labor invested in commodities tends to be ignored once they start to circulate on the market. In effect, people tend to ascribe autonomous properties or powers to commodified objects, rather than acknowledging the human labor and social relations necessary for their production. Alf Hornborg (2001) later developed the concept by applying it to the modern conception of the machine, i.e., "machine fetishism."

Industrialism   A social commitment to mass production of commodities (see Chapter 3). I use the term "advanced industrial societies" to describe industrial societies with high levels of *complexity*.

Inherently political   Following Winner (1980), this phrase is used to characterize artifacts that either necessitate or strongly encourage social relations of power.

Machine fetishism   See *fetishism*.

Materialism   A set of philosophical assertions concerning the constitution of reality as ultimately composed of matter (for an extensive explanation, see Chapter 2). This should not be confused with the common phrase "to be materialistic," which describes an attitude or set of values regarding our relation to commodities.

Matter-energy   A phrase integrating the two physical constituents of matter and energy. I often use the phrase "matter-energy throughput," which describes the quantity of matter and energy transformed by a *dissipative structure* (e.g., an economy) over time.

Metabolism   Derived from the Greek word metabolē (exchange). Denotes the material process through which organisms exchange *matter-energy* with their environment. The term was originally used by German biologists in the 1800s to explain how cells in the human body could maintain their material form over time. The term *social metabolism* denotes a socially organized material relation to the environment.

Modern/modernity   In this book, modernity refers to the social-cultural norms emerging from the Enlightenment and from the material conditions of the industrial regime (see Chapter 3).

Power density   A measure designating the horizontal land area needed per watt of energy capacity (expressed as W/m^2).

Prime mover   A device or organism transforming energy carriers into motive power that can be directed by humans to perform specific tasks.

Social metabolism   See *metabolism*.

Social-ecological   Including or referring to both social and ecological relations and processes. The term is interchangeable with "socio-ecological," "socio-environmental," and "social-environmental."

Social-ecological regime   A historically developed social-metabolic pattern of interaction between human societies and natural systems. Whereas social metabolism denotes a socially organized exchange of matter-energy with the environment, a social-ecological regime denotes the social metabolism as well as the *conceptions* necessary to support it.

Technological continuum   A concept developed in this book to analyze the social-ecological conditions of modern technologies (see Chapter 2).

Throughput   See *matter-energy*.

World economy   The current world economy is a historically developed world-system. Following Wallerstein (2011: 15), a world economy is distinguished by the economic (rather than political) linkages between its parts. This contrasts to world empires in which the system's parts share the same political unit. Many world economies have existed in history, but these have typically developed into world empires. According to Wallerstein (2004), the capitalist world-system is necessarily a world economy, since a global political unit would be likely to override the economic actors' pursuit of capital accumulation.

References

Ayres, Robert. 2016. *Energy, complexity, and wealth maximization.* Cham, Germany: Springer.

Bateson, Gregory. 2000[1972]. *Steps to an ecology of mind.* Chicago, IL and London, England: The University of Chicago Press. Bellini, Emiliano. 2020. "World's biggest PV module factory." *PV Magazine.* Published 30-03-2020. https://www.pv-magazine.com/2020/03/30/worlds-biggest-pv-module-factory/. Accessed 23-09-2020.

BP. 2020. *BP statistical review of world energy.* 69th ed. London: BP.

Cai, Mattia, Niccolò Cusumano, Arturo Lorenzoni, and Federico Pontoni. 2017. A comprehensive ex-post assessment of RES deployment in Italy: Jobs, value added and import leakages. *Energy Policy* 110: 234–245. https://doi.org/10.1016/j.enpol.2017.08.013.

de Castro, Carlos, and Iñigo Capellán-Pérez. 2020. Standard, point of use, and extended energy return on energy invested (EROI) from comprehensive material requirements of present global wind, solar, and hydro power technologies. *Energies* 13: 3036. https://doi.org/10.3390/en13123036.

DOE. 2017. *US Energy and employment report January 2017.* Washington, DC: United States Department of Energy.

DOE. 2022. *Solar manufacturing.* Office of Energy Efficiency and Renewable Energy. https://www.energy.gov/eere/solar/solar-manufacturing. Accessed 09-07-2022.

ERI. 2015. *China 2050 high renewable energy penetration scenario and roadmap study.* Beijing: Energy Research Institute National Development and Reform Commission and Energy Foundation.

Ferroni, Ferruccio, and Robert J. Hopkirk. 2016. Energy return on energy invested (ERoEI) for photovoltaic solar systems in regions of moderate insolation. *Energy Policy* 94: 336–344. https://doi.org/10.1016/j.enpol.2016.03.034.

Fraunhofer. 2020. Net installed electricity generation capacity in Germany. Updated 30-11-2020. https://energy-charts.info/charts/installed_power/chart.htm?l=enandc=DE. Accessed 02-12-2020.

Georgescu-Roegen, Nicholas. 1975. Energy and economic myths. *Southern Economic Journal* 41(3): 347–381. https://doi.org/10.2307/1056148.

GFN. 2019. Ecological footprint explorer. Published 20-09-2020. http://data.footprintnetwork.org/#/. Accessed 21-09-2020.

Gupta, Uma. 2019. Solar cells using imported blue wafers will not qualify as domestically manufactured: MNRE. *PV Magazine.* Published 22-10-2019. https://www.pv-magazine-india.com/2019/10/22/solar-cells-using-imported-blue-wafers-will-not-qualify-as-domestically-manufactured-mnre/. Accessed 23-09-2020.

Gupta, Uma. 2020. PV imports to face 20–25% customs duty in India. *PV Magazine.* Published 26-06-2020. https://www.pv-magazine.com/2020/06/26/pv-panel-imports-to-face-20-25-customs-duty-in-india-from-august/. Accessed 23-09-2020.

Haegel, Nancy M., Harry Atwater Jr., Teresa Barnes, Christian Breyer, Anthony Burrell, Yet-Ming Chiang, Stefaan de Wolf, et al. 2019. Terawatt-scale photovoltaics: Transform global energy. *Science* 364(6443): 836–838. https://doi.org/10.1126/science.aaw1845.

Hagens, Nathan J. 2020. Economics for the future—Beyond the superorganism. *Ecological Economics* 169: 106520. https://doi.org/10.1016/j.ecolecon.2019.106520.

Hornborg, Alf. 2001. *The power of the machine: Global inequalities of economy, technology, and environment.* Walnut Creek, CA: Altamira Press.

Hornborg, Alf, Gustav Cederlöf, and Andreas Roos. 2019. Has Cuba exposed the myth of "free" solar energy? Energy, space, and justice. *Environment and Planning E: Nature and Space* 2(4): 989–1008. https://doi.org/10.1177/2514848619863607.

IEA. 2017. *Snapshot of global photovoltaic markets.* Report IEA-PVPS T1-31:2017. Paris, France: Photovoltaic Power Systems Programme of the International Energy Agency.

IEA. 2020a. *Total energy supply (TES) by source, China (People's Republic of China and Hong Kong China) 1990-2018.* Published 06-2019. https://www.iea.org/data-and-statistics?country=CHINAREGandfuel=Energy%20supplyandindicator=TPESbySource. Accessed 09-12-2020.

IEA. 2020b. *Total energy supply (TES) by source, Germany 1990-2019.* Published 04-06-2019. https://www.iea.org/data-and-statistics?country=GERMANYandfuel=Energy%20supplyandindicator=TPESbySource. Accessed 09-12-2020.

IEA. 2020c. *Total energy supply (TES) by source, India 1990-2018.* Published 04-06-2019. https://www.iea.org/data-and-statistics?country=INDIAandfuel=Energy%20supply-andindicator=TPESbySource. Accessed 09-12-2020.

IEA. 2020d. *Total energy supply (TES) by source, Italy 1990-2019.* Published 04-06-2019. https://www.iea.org/data-and-statistics?country=ITALYandfuel=Energy%20supply-andindicator=TPESbySource. Accessed 09-12-2020.

IFC. 2015. *Utility-scale solar photovoltaic power plants: A project developer's guide.* Washington, DC: International Finance Corporations.

IRENA. 2017. *Renewable energy and jobs—Annual review 2017.* Abu Dhabi, UAE: International Renewable Energy Agency.

Llera, E., S. Scarpellini, A. Arandra, and I. Zabalza. 2013. Forecasting job creation from renewable energy deployment through a value-chain approach. *Renewable and Sustainable Energy Reviews* 21: 262–271. https://doi.org/10.1016/j.rser.2012.12.053.

Marx, Karl. 1990[1867]. *Capital: A critique of political economy.* London, England: Penguin.

MERCOM. 2020. Q1 2020 India solar market update—1,080 MW installed in Q1 2020. *MERCOM India.* https://mercomindia.com/product/q1-2020-india-solar-market-update/

NECP. 2019. Integrated national energy and climate plan. *Ministry of Economic Development, Ministry of the Environment and Protection of Natural Resources and the Sea and Ministry of Infrastructure and Transport.* https://climate-laws.org/geographies/poland/policies/integrated-national-energy-and-climate-plan-2021-2030#:~:text=The%20National%20Energy%20and%20Climate,overall%20greenhouse%20gases%20emissions%20targets

Nemet, Gregory. 2019. *How solar energy became cheap: A model for low-carbon innovation.* London, England and New York, NY: Routledge.

OECD. 2019. Average annual hours actually worked per worker: Total employment. *OECD Stat.* https://stats.oecd.org/Index.aspx?DataSetCode=ANHRS#. Accessed 20-09-2020.

Prieto, Pedro A., and Charles A. S. Hall. 2013. *Spain's photovoltaic revolution: The energy return on investment*. New York, NY: Springer.

Ritchie, Hannah, and Max Roser. 2020. United States: Energy country profile. *Our World in Data*. Published 06-07-2020. https://ourworldindata.org/energy/country/united-states. Accessed 07-07-2022.

Spencer, Thomas, Neshwin Rodrigues, Raghav Pachouri, Shubham Thakre, and G. Renjith. 2020. *Renewable power pathways: Modelling the integration of wind and solar in India by 2030*. New Delhi, India: The Energy and Resources Institute.

Sultan, Zafar Ahmad, and Tarek Tawfik Yousfe Alkhateeb. 2019. Energy consumption and economic growth: The evidence from India. *International Journal of Energy Economics and Policy* 9(5): 142–147.

Tainter, Joseph A. 1988. *Collapse of complex societies*. Cambridge: Cambridge University Press.

Terzini, E., A. De Lillo, F. Di Mario, C. Privato, G. Graditi, G. Di Francia, A. Antonaia, A. Mittiga, P. Delli Veneri, and C. Minarini, eds. 2011. Quaderno: Fotovoltaico. Agencia nazionale per le nuove technologie, l'energia e lo sviluppo economic sostenibile.

Trading Economics. 2020. India GDP. *Trading Economics*. https://tradingeconomics.com/india/gdp. Accessed 09-12-2020.

Trend Economy. 2020. https://trendeconomy.com/. Accessed 14-04-2020.

UN Comtrade Database. 2021. https://comtrade.un.org/data/. Accessed 03-2021.

Wackernagel, Mathis, and Chad Monfreda. 2004. Ecological footprints and energy. *Encyclopedia of Energy* 2: 1–11.

Wallerstein, Immanuel. 2004. *World-systems analysis: An introduction*. Durham, NC and London, England: Duke University Press.

Wallerstein, Immanuel. 2011. *The modern world system I: Capitalist agriculture and the origins of the European world-economy in the sixteenth century*. Berkeley, Los Angeles and London, England: University of California Press.

Warr, Benjamin, and Robert Ayres. 2010. Evidence of causality between the quantity and quality of energy consumption and economic growth. *Energy* 35(4): 1688–1693. https://doi.org/10.1016/j.energy.2009.12.017.

Winner, Langdon. 1980. Do artifacts have politics? *Daedalus* 109(1): 121–136.

Worldometer. 2020a. China GDP. https://www.worldometers.info/gdp/china-gdp/. Accessed 09-12-2020.

Worldometer. 2020b. Germany GDP. https://www.worldometers.info/gdp/germany-gdp/. Accessed 09-12-2020.

Worldometer. 2020c. Italy GDP. https://www.worldometers.info/gdp/italy-gdp/. Accessed 09-12-2020.

Index

Note: **Bold** page numbers refer to tables; *italic* page numbers refer to figures and page numbers followed by "n" denote endnotes.

For Product Safety Concerns and Information please contact our EU
representative GPSR@taylorandfrancis.com
Taylor & Francis Verlag GmbH, Kaufingerstraße 24, 80331 München, Germany

www.ingramcontent.com/pod-product-compliance
Lightning Source LLC
Chambersburg PA
CBHW060256220326
41598CB00027B/4126